The Physics of Sound Waves (Second Edition)

Music, instruments, and sound equipment

The Physics of Sound Waves
(Second Edition)

Music, instruments, and sound equipment

Panos Photinos

Southern Oregon University, Ashland, OR, USA

IOP Publishing, Bristol, UK

Multimedia content is available for this book from http://iopscience.iop.org/book/978-0-7503-3539-3.

ISBN   978-0-7503-3539-3 (ebook)
ISBN   978-0-7503-3537-9 (print)
ISBN   978-0-7503-3540-9 (myPrint)
ISBN   978-0-7503-3538-6 (mobi)

DOI   10.1088/978-0-7503-3539-3

Version: 20210701

IOP ebooks

British Library Cataloguing-in-Publication Data: A catalogue record for this book is available from the British Library.

Published by IOP Publishing, wholly owned by The Institute of Physics, London

IOP Publishing, Temple Circus, Temple Way, Bristol, BS1 6HG, UK

US Office: IOP Publishing, Inc., 190 North Independence Mall West, Suite 601, Philadelphia, PA 19106, USA

Cover image: Gift of Mary Phelps Smith, in memory of her husband, Howard Caswell Smith, 1965 (Metmuseum.org) (Public Domain).

To my mother Irini, to Demetri, Marko, Ryan, Zoe, and to Shelley who made it all possible.

Contents

Preface

I had the good fortune to grow up without TV. My family's entertainment was radio, records, playing instruments and singing. My mother had a beautiful warm voice. I took piano lessons early on, from a lovely Austrian teacher, Madame Anselmi. I never knew her first name, because in my culture at the time, asking your teacher's first name was not proper. As a teenager, I helped at my parents' sound-equipment store, and was soon able to set up sound equipment for band performances. Choosing a major was a great dilemma. I ended up in physics rather than music. A dual major was not an option at the University of Athens.

I had the opportunity to develop courses for non-science majors, and thoroughly enjoyed teaching them. This book is based on my lecture notes on the subject. My experience in teaching these courses taught me to get to the music subject as early as possible, and so, in this book I present science and music right from the first chapter, rather than doing all the science first, and then the music.

In writing this book, I avoided the use of equations in the text, as much as possible. I include many examples, questions and problems, with detailed solutions in the appendix. I also included videos and audio, to amplify the descriptions in the text. The book is intended for general education courses and for non-science majors, but should be very useful for amateur musicians, and a general readership as well.

As for me following a career in physics instead of my passion for playing and composing music, I now realize that music is a lifelong companion, and that a musical instrument is a most receptive confidante.

Acknowledgements

It is a pleasure to acknowledge Ashley Gasque, Robert Trevelyan and Chris Benson of IOP Publishing for their guidance and expert help in preparing this book. I thank Dr Demeter Tsounis, Dr Petros Vlachopoulos, Mr Wataru Sugiyama, and Ms Linda Chambers for sharing their knowledge and music performing expertise, and Dr Gordon Wolfe for teaching me how to appreciate many genres of music. I am thankful to Mr Michali and Mr Aleko Ieronymidi for their friendship and for sharing their expertise on instrument making. I very much appreciate discussions I had with Mr Liran Ozeri on sound therapy, Ms Sharon Narduzzi on native American instruments, and Janet Bocast for help with the art work. I am grateful to my wife Shelley for her editing and constant encouragement.

Author biography

Panos Photinos

 Panos Photinos is professor emeritus at Southern Oregon University where he has taught since 1989. He developed and taught many courses, including two courses on the physics of music. Prior to joining SOU he held faculty appointments at the Liquid Crystal Institute, Kent, Ohio; St Francis Xavier, Antigonish, Nova Scotia, Canada; and the University of Pittsburgh, Pennsylvania. He was visiting faculty at the University of Sao Paulo, Brazil, the University of Patras, Greece, and Victoria University in Wellington, New Zealand. Panos completed his undergraduate degree in physics at the National University of Athens, Greece, and received his doctorate in physics from Kent State University, Ohio. He started piano lessons at the age of five in Egypt, where he was exposed to a wide range of musical traditions, including Arabic, Armenian, Berber, Indian, Jewish, and Turkish.

During his college years in Athens, he supplemented his income playing piano and guitar at various nightclubs in the famous district of Plaka, at the foothill of the Acropolis. He is a collector of traditional musical instruments. He enjoys music sessions with his family in Ashland, Oregon, and with his relatives in South Australia and from his homeland, the island of Ikaria, Greece. Panos has authored over 50 research publications in scientific journals, and is the author of *Visual Astronomy: A guide to understanding the night sky*.

IOP Publishing

The Physics of Sound Waves (Second Edition)
Music, instruments, and sound equipment
Panos Photinos

Chapter 1

Properties of waves

1.1 Introduction

In everyday language the term *wave* has several uses; for example, a wave of e-mails, a wave of enthusiasm, a wave of applause, a heat wave, and so on. The general idea is that something is suddenly going above or below normal level. In more technical language, a wave indicates cycles of highs and lows in some quantity. In terms of sound, what is 'waving' is the air pressure, going higher and lower than the ambient atmospheric pressure. This chapter will introduce the basic concepts that are commonly used to characterize waves, and sound waves in particular.

doi:10.1088/978-0-7503-3539-3ch1

1.2 Periodic waves

A most familiar wave is the pattern of circles generated by dropping a coin in a still pond, shown in figure 1.1. The pattern consists of highs and lows (**crests** and **troughs**, respectively) traveling outwards from the center. At each point of the surface of the pond, the water level **oscillates** in cycles, above and below the undisturbed level of the pond. The undisturbed level is the **equilibrium** level of the water surface. The pattern propagates away from the point of impact (the source of the wave) and at each point the propagation is along the line of sight to the source. In a shallow flat-bottomed pond, the crests (and troughs) travel roughly at the same speed. As the speed is the same for all crests, the distance between successive crests remains the same as the wave travels. The distance between successive crests (or successive troughs) is defined as the **wavelength**. The time elapsed between the crossings of two successive crests through a given point is the **period** of the wave. In one period, the wave travels a distance of one wavelength; in other words, the **speed** (which is distance traveled divided by time of travel) equals the ratio of the wavelength divided by the period. In air, sound waves travel at a speed of about 343 meters per second.

If we count the number of crests crossing through one point in a given time interval, say in one second, then we have a very important concept in the study of sound, which is the **frequency**. The frequency is equal to the inverse of the period, that is:

$$\text{frequency} = 1/(\text{period}) \quad \text{and therefore} \quad \text{period} = 1/(\text{frequency}).$$

If four successive crests cross a given point in 1 s (that is if the frequency is 4 crests per second) then the time elapsed between two successive crests, that is the period, is 1/4 of a second. As the speed of the wave equals the wavelength divided by the period, and since frequency is the inverse of the period, it follows that the speed of the wave is equal to the product of the wavelength times the frequency:

$$\text{Speed of wave} = (\text{frequency}) \times (\text{wavelength}).$$

The frequency is fundamental in characterizing sound tones, and is a measure of what we call the pitch. High pitch tones correspond to high frequencies, and low pitch tones correspond to low frequencies. The frequency is measured in units of Hertz (**Hz** for short).

Figure 1.1. A wave in a pond.

The difference in height between the top of a wave crest and the undisturbed water level is the **amplitude** of the wave. The amplitude of the wave depends on the weight and speed of the impacting object. A small coin will create a wave of smaller amplitude than a huge rock. The amplitude relates to the intensity of the wave, or the energy carried by the wave. Note that as the wave spreads out the amplitude diminishes, and eventually the wave dies out. With sound, this observation relates to everyday experience: the farther we are from the source, the weaker it sounds.

The fact that the water level moves up and down, means that there must be a force involved, which we will call the **restoring force**. In the opening figure of this chapter, we see small ripples. In this case the restoring force comes mainly from the surface tension of water[1]. In the waves shown in figure 1.1, the force involved is gravity. In either case, water that happens to be above the equilibrium level of the surface is pulled down by the restoring force. The downward speed builds up, and that amount of water falls below the equilibrium level of the surface and becomes part of a trough. While moving down it pushes adjacent parts of the water upward, which become part of a crest, and so forth. We can think of the entire cycle as an attempt by the restoring force to bring the water level back to equilibrium. In the process of restoring the equilibrium water level, the water surface keeps overshooting the target, which is the equilibrium level.

All waves require a restoring force. For example, in a vibrating string, the tension of the string acts as the restoring force. There is a relation between the frequency of the wave and the strength of the restoring force. A *stronger restoring force* makes the up-and-down oscillation faster, meaning that the *frequency will be higher*. This relation will be discussed in more detail in connection with strings and string instruments.

In the example of the water wave, the quantity that oscillates is the water level, as compared to the equilibrium level, in other words the oscillating quantity is a *distance*, measured in meters, inches and so on. In the case of sound waves in air, the oscillating quantity is the air *pressure*, which is commonly measured in Pascals or in psi. The sound wave in air is a succession of layers of low and high pressures, the **rarefactions** and **compressions**, respectively. Low and high pressures are with reference to the undisturbed air pressure of the surrounding atmosphere. Note that, in the pond example, as the wave travels in the horizontal direction, the oscillation is up and down. In other words, the oscillation is at a right angle to the direction of travel. This is an example of a **transverse wave**: the oscillation is transverse to the direction of travel. In the case of a sound wave in air, the pressure oscillates back and forth, along the direction of travel. This is an example of a **longitudinal wave**.

The medium in which the wave is travelling does not necessarily travel with the wave. For example, if we stretch a long rope between two posts, and pluck it at one

---

[1] For small *ripples*, that is wavelengths shorter than about 10 cm (4 inches), the surface tension of the water becomes the dominant restoring force. Surface tension is a manifestation of the cohesiveness of water which opposes any deformation of the shape of the water surface. For instance, surface tension allows insects to stand on their legs on top of the water surface rather than sinking in.

end, a transverse wave will travel along the rope, but the rope itself will stay in place. In the same way, in our pond example, a leaf or an insect floating in a pond will not travel with the crests or troughs of the wave. Some waves do not require a medium at all, because they can travel through empty space (see section 1.6.)

It is convenient to use graphs to represent waves. The simplest periodic graph is the **sinusoid**, meaning a graph described by the *sin* (sine) or *cos* (cosine) functions familiar from trigonometry. We will refer to these graphs as **waveforms**. Three cycles of a sinusoidal wave form are shown in figure 1.2. The cycle or **periodicity** of the waveform is equal to the distance between two *successive* highs or two successive lows. The **phase** at any point of the waveform is the fraction of the cycle elapsed from the starting point of the graph. For example, in figure 1.2 the phase at the first high on the left is one-quarter of a cycle; the phase at the first low is three-quarters of a cycle.

When using graphs to represent waves, the vertical axis of the graph is used to show the value of the oscillating quantity, which depending on the nature of the wave described, could be displacement, pressure, and so on. For the horizontal axis we usually have two options. We can choose the horizontal axis to show distance or time, as indicated in figure 1.3.

If we use figure 1.3 to describe the water wave in a pond, then the vertical axis is the height of the water level. From the graph, we see that the amplitude of the wave is 1 m. If we choose to show distance in the horizontal axis, then the graph will give the profile of the height at a given instant, like a snapshot of the wave. If we choose the horizontal axis to indicate time, then the graph shows how the height of the water level evolves with time, at a *given point*. So, the two graphs in figure 1.3 give us different information about the wave. If we recall the definitions of the wavelength and period, we see that the distance between successive crests in figure 1.3(a) is equal to the wavelength. In this example, the first crest occurs at a distance of about 1.8 m from the '0' of the horizontal axis, and the second crest occurs at a distance of about 7.8 m. So, we can find the wavelength by taking the difference of the locations of the two crests, that is 7.8 − 1.8 = 6 m. In figure 1.3(b), the first crest occurs at time = 1 s, and the second crest at time = 5 s. The time interval between successive crests is the period of the wave, and in this example, it is 5−1 = 4 s. As the frequency is the inverse of the period, it follows that the frequency of this waveform is (1/4) = 0.25 Hz. We will use the term **periodicity** to refer to the extent of the cycle. If the horizontal axis

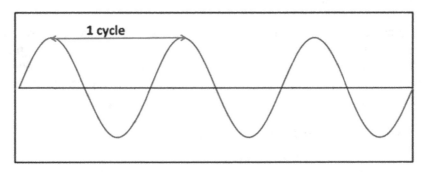

Figure 1.2. Three cycles of a sinusoidal waveform.

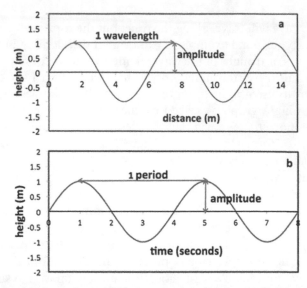

Figure 1.3. Sinusoidal waveform. (a) Horizontal axis is distance. (b) Horizontal axis is time.

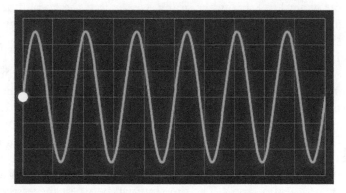

Video 1.1. Wave travelling from left to right. The yellow dot oscillates up and down at the frequency of the wave. The wavelength equals the distance between successive crests. Video available at https://iopscience.iop.org/book/978-0-7503-3539-3.

is distance, the periodicity means wavelength. If the horizontal axis is time, then periodicity means period.

To get a more complete picture of the time evolution of the wave, one would need either a series of snapshots like figure 1.3(a), or a series of height versus time, like figure 1.3(b) at *different points* along the path of the wave. Video 1.1 shows a time sequence of snapshots of the wave. The horizontal axis represents the distance from the origin of the wave.

1.3 Addition of waveforms

In this section we discuss simple ways in which waves can combine with each other. Comparison of the phase of the interacting waves is the key concept in understanding the outcome. Figure 1.4 shows two identical waveforms. The horizontal

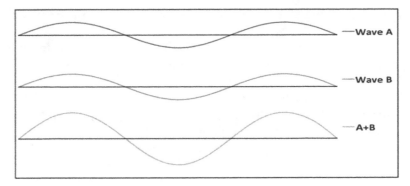

Figure 1.4. Adding two identical waveforms that are in-phase.

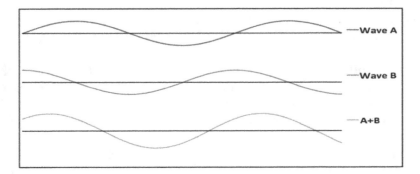

Figure 1.5. Adding two identical waveforms that are 1/4 of a cycle out-of-phase.

axis is not labeled, and can be either distance or time without affecting the conclusions. The graphs are offset vertically, and the horizontal lines represent zero displacement for each wave. To find the waveform resulting from combining waves A and B, we add the displacements at each point of the horizontal axis. The result is shown in the bottom graph, and is simply the sum of the two waves, that is, the amplitude of the resulting wave is doubled, and the periodicity (which is the wavelength or the period depending on the choice of the horizontal axis) remains the same. In this case, we combined two waves that are in step, or **in-phase**. This means at each point of the horizontal axis, the two waves have the same phase, and the **phase difference** between them is zero.

In figure 1.5, wave B is displaced to the left by one quarter of a cycle. In this case the two waves are not in-step. The two waves are **out-of-phase** by one-quarter of a cycle. If the horizontal axis were distance, the crests of the two waves would be separated by one quarter of a wavelength. In the same way, if the horizontal axis indicated time, then the crests of the two waves would be separated by one quarter of the period. To find the waveform resulting from combining waves A and B, we add the displacements at each point of the horizontal axis. The result is shown in the bottom of figure 1.5. We note that the sum of waves A and B has the same periodicity, but the amplitude is smaller than the sum of the amplitudes of A and B.

The crest of the resulting wave occurs somewhere in between the crests of waves A and B.

In figure 1.6 wave B is displaced to the left by half a cycle, and the result of adding these two waves is total cancellation. As the amplitudes of A and B are the same and the oscillations are in opposite directions, the cancellation is complete. If the amplitudes were not the same, the cancelation would not be complete, and the resulting wave would have amplitude equal to the difference between the amplitudes of A and B. For instance, if A had amplitude 3 and B had amplitude 1, the resulting wave would have amplitude equal to the difference $3 - 1 = 2$.

Using similar diagrams, we can add waves of the same periodicity. The general conclusions are:
- The resulting wave will always have the same wavelength and frequency.
- The amplitude will be equal to the sum of the amplitudes, if the phase difference is zero.
- The amplitude will be equal to the difference between amplitudes, if the phase difference is 1/2 cycle.

If the phase difference between waves A and B has value between 0 and 1/2 cycle, the value of the amplitude of the resulting waveform will be somewhere between the sum of amplitudes (A + B) and the difference between amplitudes (A—B) or (B − A) if the amplitude of B is larger, because by convention, the amplitude is always a positive number. Note that in our discussion we applied the concept of phase difference to waves of the *same* frequency, and wavelength. Also, note that in some cases we will use degrees rather than cycles. For instance, we would say that in figure 1.5 the two waves are 90° out of phase, and in figure 1.6 we would say that the two waves are 180° out of phase.

Cancellation of waves as shown in figure 1.6 is the basic principle behind **noise cancelling** devices. For example, noise cancellation headphones use electronics to generate replicas of the sound before it reaches each ear. The replicas generated differ in phase from the incoming wave by 180°, meaning that they are opposite in sign, so the incoming sound and its out-of-phase replica cancel each other.

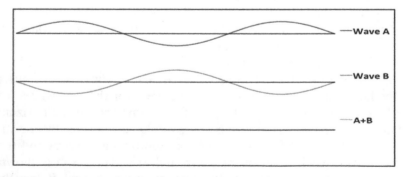

Figure 1.6. Adding two identical waveforms that are 1/2 cycle out of phase. The result is complete cancellation.

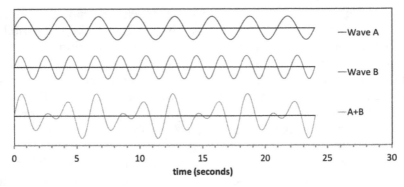

Figure 1.7. Adding two waves of the same amplitude and different frequencies. Wave A completes 2 for every 3 cycles of wave B. The result (A+B) is not a simple sinusoid, and completes a full cycle after Wave A completes 2 cycles.

For the moment, let us assume that the horizontal axis in figures 1.4–1.6 is time. So, the graphs represent waveforms of the same frequency. One may ask: what is the effect of phase when adding waveforms of different frequencies? In terms of musical sound, the answer is that the effects of phase are of minor significance, so in what follows we will discuss the result of adding waves of different frequencies and ignore the phase difference. For our discussion we will use very low frequencies to keep the calculations simple[2].

Sound waves, even single notes from a musical instrument, are almost always combinations of different frequencies. Of particular interest to musical sound are combinations of pairs of frequencies that are related by a ratio of small numbers, for example 3:2, 4:3 and so on. As before, to get the resulting waveform, we add the values of the displacement at each point of the horizontal axis.

Figure 1.7 shows two waveforms of equal amplitude. The two waveforms have periods 3 s (Wave A) and 2 s (Wave B), respectively. The displacement of the combined waveform reaches as high as double the amplitude of the individual waves, and remains lower in the in-betweens. The ratio of the two periods is 3:2. In other words, by the time A completes 2 full cycles, B completes 3 full cycles. It takes $2 \times 3 = 6$ s for Wave A to complete 2 cycles. Also, it takes $3 \times 2 = 6$ s for Wave B to complete 3 cycles. Either way, the combined waveform repeats every 6 s. We note that the result is no longer a simple sinusoid. The pattern repeats at a rate of $1/6 = 0.166$ Hz.

Example 1: We will add two sinusoids with periods 3 and 4 s. In this case the frequency ratio is 4:3.

The combined waveform has a repeat period of $3 \times 4 = 12$, and the pattern repeats at a rate of $1/12 = 0.068\ 33$ Hz.

Note that a combining two waves of frequency ratio 4:3 means the same thing as combining two waves of frequency ratio 3:4. The same applies for the period ratios.

---

[2] The mathematical equations for addition of waves are given in appendix B.3.

Note also that the ratio 3:2 is the same as 6:4 or 12:8. Here we have common factors, and in this case, we work with the smaller numbers, 3:2, so the repeat period is 3×2 rather than 4×6 or 8×12 s. In terms of musical sound, it is important to note that the key quantity is the ratio of the frequencies of the two waves we combine. The ratio of frequencies is called an interval, and the higher frequency is in the numerator. In general, if we are combining two waves of periods T_1 and T_2, and ratio $n{:}m = T_1{:}T_2$ where n and m are integers with no common factors, then the combined waveform repeats with period $n \times T_2$ or $m \times T_1$.

Example 2: Combine two waves A and B, of periods of 0.003 and 0.002 s.

First, we find the ratio $0.003 : 0.002 = 3{:}2$. So, the combined wave repeats with period 2×0.003 s $= 0.006$ s, (which is the same thing as 3×0.002.) This means that the pattern repeats at a rate of $1/0.006 = 166.7$ Hz.

Note that since the combination is not a 'pure' sinusoid, we cannot assign a single frequency to it. This lack of 'purity' of the combined wave by no means implies a lower quality of sound. Indeed, the two frequency combinations 3:2 and 4:3 are the so-called *perfect fifth* and *perfect fourth*, respectively, and as will see, these particular combinations, as indicated by their name, are most pleasing to the ear[3]. What is not pleasing to the ear is the result of combining two waves of almost the same frequency, which we will discuss next.

1.4 Beats

A very interesting situation occurs as the frequencies of the two waveforms get closer. Figure 1.8 shows two waves, A and B, of equal amplitude, but the frequencies now are closer to each other than in the previous examples. Again, we see that the displacement reaches as high as double the amplitude of the individual waves, and remains lower in the in-betweens. By counting crests, we see that in the time it takes

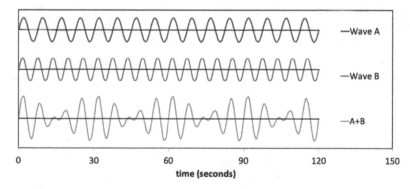

Figure 1.8. Adding two waveforms of the same amplitude with frequencies in the ratio 5:4.

---

[3] See section 5.6.

wave A to complete 4 cycles, wave B completes 5 cycles. In other words, the period ratio of the two waves is 5:4. This ratio corresponds to the so-called major third. The actual periods in this example are 7.5 and 6 s and the repeat period is $4 \times 7.5 = 30$ s. So, the combined waveform repeats with frequency $1/30 = 0.0333$ Hz.

In the next example we will make the frequencies even closer. We will combine two waveforms of frequencies 10 and 10.6 Hz. The frequency ratio in this case is $10.6:10 = 1.06$, a value which is very close to the ratio of two consecutive notes on a piano keyboard[4]. In terms of full numbers, the frequency ratio is 106:100. Figure 1.9 shows the result. The value of the waveform gets fairly low every 1.66 s but actually it does not repeat exactly! Note that the wiggles at about 0.9 and 2.5 s, are not the same. Looking at this waveform as a sound wave, one would notice a regular change in the 'volume' every 1.66 s, meaning a recurring variation in the volume at a frequency of $1/0.166 = 0.6$ Hz. What is interesting here is that this frequency is the difference between the frequencies of the two waveforms. One could scale up the frequencies to more practical values for musical tones, as is done in the next example.

Example 3: Combine two tones with frequencies 439 and 441 Hz.

In this case, the combined waveform repeats with frequency equal to the difference of the individual frequencies, that is $441 - 439 = 2$ Hz. As before, the displacement of the combined waveform reaches highs equal to double the amplitude of the individual waveforms, and remains low in the in-betweens. These highs and lows repeat twice every second (2 Hz) which is slow enough for the human ear to pick up. So, the amplitude of the resulting waveform is **modulated**, that is, it changes with time following a regular pattern. This periodic modulation of the amplitude is referred to as **beats**, and is generally unpleasant to the ear. In example 3, the modulation repeats with a frequency of 2 Hz. This is the **beat frequency**, meaning that the highs of the amplitude will repeat every $1/2 = 0.5$ s. In this example, the two frequencies are too close, and our ear cannot tell them apart. What we perceive is a sound frequency equal

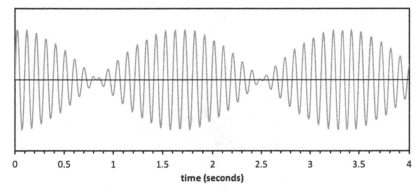

Figure 1.9. Sum of two waves of frequencies 10 and 10.6 Hz.

---

[4] As we will see in section 5.5 the actual ratio for two successive keys is 1.059 46.

to the average of the individual frequencies, which is $(441 + 339)/2 = 440$ Hz, the so-called **carrier frequency**, with a sound volume that cycles every 0.5 s.

The phenomenon of beats plays a significant role in music and musical instruments, particularly those with double or triple strings, such as the piano. Suppose that we have two strings, one vibrating with frequency 441 Hz, and the other vibrating with frequency 339 Hz. What we will hear is a frequency of about 440 Hz, and the intensity of the sound will fluctuate up and down every 0.5 s, producing a tone of poor quality.

Figure 1.10 shows the waveform of the sound produced by sounding two[5] strings of the C5 note on a piano. The entire waveform lasts about 2 s, and shows about 15 beats, so the beat frequency is about 7.5 Hz, which tells us that the two strings are off by 7.5 Hz. The beat pattern is captured in Audio 1.1. As will be discussed in section 5.6 this is quite noticeable and unpleasant to the ear. Experienced tuners and string instrument players can use beats to tune their instruments very precisely.

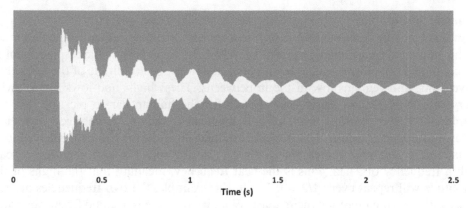

Figure 1.10. Sound produced by an out-of-tune piano, showing 15 beats in total over a period of about 2 s, or 7.5 beats per second. The pitch of the note used is 532 Hz.

Audio 1.1. Sound of waveform shown in figure 1.10. Available at https://iopscience.iop.org/book/978-0-7503-3539-3.

---

[5] One of the three strings of the C5 note was muted.

1.5 Energy and intensity

Waves carry energy, which is measured in **Joules** (J for short.) So, when we drop a coin in a pond, some of the energy of the falling coin (the so-called kinetic energy) is transferred to the water, and carried away by the wave generated by the impact. A heavier coin will result in a wave of larger amplitude, because it imparts a larger amount of energy at the point of impact. In many cases, the important quantity is not the energy imparted at the point of impact, but how much energy is *flowing* away from the source of the wave. To understand the concept of energy flow, it would be helpful to make an analogy to collecting rainwater in a tub. The more rain is coming down, the more raindrops will be caught in the tub, but the amount of water collected will also depend on the size of the tub because a wider tub has larger collecting area. The amount of water collected will depend on time as well: the longer we wait, the more water will be collected. To characterize the intensity of the rainfall, we need to specify the tub area (say 1 m^2) and the time interval (say in 24 h). Also note that the amount collected will depend on the tilt of the tub. A vertical tub will not catch much water! The maximum amount of water will be collected if the tub is perpendicular to the direction of the falling raindrops.

The *rate* of energy flow (that is, the amount of energy flowing per unit time) is defined as the **power**, and it is measured in **Watts** (W). For waves, we can define the **intensity** as the *rate* at which energy is flowing through an area of 1 m^2. The area must be oriented perpendicular to the direction of energy flow. Therefore, the intensity is measured in Watts per square meter (W m^{-2}). The intensity is related to the square of the amplitude of the wave. This means that if we double the amplitude of the wave the intensity is not doubled but quadrupled (following the square of 2, that is $2^2 = 4$). If the amplitude is tripled, the intensity increases $3^2 = 9$ times, and so on. The nature of the amplitude depends on the type of wave in question. For water waves, the amplitude refers to the difference between the equilibrium level of the water and the height of a crest. The same for a vibrating string; the amplitude refers to the maximum displacement of the string from the equilibrium position. If we are talking about a sound wave in air, then the amplitude refers to the difference between the atmospheric pressure and the pressure of the sound wave at its max. This difference is what we call the acoustic pressure, or acoustic wave, which will be discussed further in section 2.3

As the wave propagates outward from the source, the energy that started at the source keeps spreading out over a larger and larger surface. The result is that the intensity (and amplitude) will decrease more and more the further we move away from the source. In addition, some of the energy that started at the source is absorbed, usually converted to heat as a result of friction. We refer generically to the decrease in intensity as **attenuation**. Attenuation will be discussed in detail in section 2.7.

1.6 Further discussion

Examples of waves

Electromagnetic (EM) waves include light, radio waves, microwaves, ultraviolet radiation, x-rays, and more. The oscillating quantity in EM waves is an electric field combined with a magnetic field. EM waves do not require a medium to propagate. This is why we can receive light and other EM waves from stars and galaxies. The EM waves in this case travel huge distances through space, which is an almost perfect vacuum, before reaching us. In empty space, all EM waves travel at the speed of light. EM waves can be produced by the oscillation of electrical charges. This is exactly what happens in a radio or TV antenna. Electric changes are moving back and forth creating EM waves.

WiFi is a wireless connection between devices based on transmitting and receiving EM waves of frequency 2.4 billion Hz (GHz for short) and 5 GHz. Similarly, an **AM radio** is a wireless connection, using EM waves of frequency in the range of 540–1600 thousand Hz (kHz). **FM radio** uses EM waves of frequency in the range from 87.5 to 108 million Hz (MHz). All these waves carry the sound information, but what we hear is not the EM wave itself. The EM wave is only the carrier of information, the so-called **carrier wave**. The sound information is 'imprinted' on the carrier signal as a **modulation** of the amplitude (AM = **A**mplitude **M**odulation) or the frequency (FM = **F**requency **M**odulation) of the carrier wave. The waveform in figure 1.9 is an amplitude-modulated signal.

Tsunamis are series of waves usually created by underwater earthquakes or volcanic explosions. As the waves approach the shore, their height increases reaching about 10 m, with crests separated about 10 km, moving at a speed of about 50 km h^{-1}. Ocean waves approaching the coast can have similar speeds and heights, but the wavelength of tsunami waves is thousands of times larger. In other words, the high level of water during the crest of the tsunami wave lasts thousands of times longer than it does in ordinary waves, creating a flood of catastrophic proportions.

One of the most exciting discoveries in recent years was the detection of **gravitational waves**. In principle, gravitational waves can be produced by pairs of massive stars that move around each other in a periodic way. The oscillating quantity in the case of gravitational waves is the 'fabric' (or the curvature) of space, as predicted by Einstein's general theory of relativity.

Brain waves is a term used to describe minute electric voltage variations on the human sculp. These voltage variations are rhythmical and can include frequencies ranging from about 1 Hz up to about 100 Hz. They are measured using electro-encephalography (EEG for short) which is a technique similar to the EKG which is used to measure electrical signals associated with the function of the heart. The signals measured by EEG on the scalp reflect electrical currents inside the brain. These currents are caused by synchronized activity of large groups of brain cells. So essentially brain waves are electromagnetic in nature. What is interesting is that there is a direct correlation between the frequency of the brain wave and the level of the brain's activity. Roughly speaking, frequencies below 4 Hz (the so-called delta band) occur during sleep. Slightly higher frequencies correspond to a very relaxed state, while at the higher end (about 30 Hz and higher) the brain is most alert and focused.

1.7 Equations

The following symbols are commonly used:

c = speed

λ = wavelength

f = frequency

T = period

f is defined as $f = 1/T$

For sinusoidal waves we have:

$c = f\lambda$ and by the definition of f we have $c = \lambda/T$.

1.8 Problems and questions

1. A person's heart rate is 120 beats per minute.
   (a) Find the period of the heart rate in seconds.
   (b) Find the frequency of the heart rate in Hz.
2. A wave has frequency 20 Hz and wavelength 200 m. Find the speed of the wave.
3. Suppose we have two sound waves, A and B, traveling in air simultaneously. The amplitude of wave B is twice the amplitude of wave A.
   (a) Which of the two waves has higher intensity?
   (b) Which of the two waves travels faster?
4. Suppose we combine two sound waves of frequencies 104.3 and 100.1 Hz. Find the beat frequency resulting from the combination of the two waves.
5. Suppose that we combine two waveforms of periods 2 and 7 Hz. How often does the pattern of the combined waveform repeats?
6. This exercise uses video 1.1.
   (a) What is the wavelength of the sinusoid shown in video 1.1? Assume that the vertical gridlines are 1 m apart.
   (b) How long does it take for the yellow dot to complete a full oscillation cycle? (You will get more accurate results if you measure the time it takes to complete 6 cycles and then divide the time by 6.)
   (c) What is the frequency of the wave?
   (d) What does the product of the wavelength times the frequency represent?
   (e) Find the speed of the wave by measuring the time it takes for one crest to travel a distance of 5 m. How does your measurement compare to the result in part (d)? In this part you should use your own timer, because the video timer has low resolution.

Chapter 2

Propagation of sound waves

Echo by Alexandre Cabanel, The Metropolitan Museum of Art (http://www.met-museum.org/art/collection/search/435829).

2.1 Introduction

In chapter 1, we introduced the general properties of waves, such as wavelength, amplitude, speed and so on. In this chapter, we will consider how these properties are affected by the medium and the environment in which the wave propagates. For instance, in closed spaces a sound wave will encounter obstacles such as walls, furniture, and other. In the outdoors, besides obstacles (ground, buildings, trees and so on) the wave is affected mainly by changes in the temperature of the atmosphere. The general behavior of the various types of waves can be similar in many cases; for example, the path of sound waves can be bent if the air temperature along their path changes, which is what also happens to light waves. But there are significant differences as well. For instance, one can block light by placing one's hand in front of the light source. This would not work with sound! The wavelength is the key to understanding any differences in the behavior of sound compared to light. Since behavior of light rays is easier to visualize, we will occasionally refer to the behavior of light for comparison.

2.2 Wave fronts

In order to describe the propagation of waves, it is useful to introduce the concept of **wave fronts**, which is closely associated with the concept of phase, introduced in section 1.2. We refer to the familiar example of the circular pattern created by dropping a coin in a pond, represented schematically in figure 2.1. All the crests belonging to a given circle have the same phase; in other words, they have completed

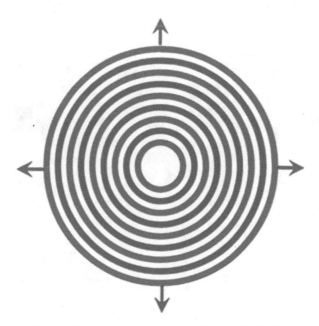

Figure 2.1. Circular wave fronts. Blue circles indicate crests and red circles indicate troughs. Adjacent crests are separated by one wavelength. The same applies to adjacent troughs. The arrows indicate the direction of propagation of the wave fronts.

the same fraction of their cycle. The arrows indicate the direction of propagation. Assume that the blue circles represent crests and the red circles represent troughs. Compared to a given blue circle, the crests on the blue circle ahead of it have completed one cycle more, meaning that they are one wavelength ahead. The crests on the blue circle behind it are one wavelength behind. In other words, the distance between adjacent blue circles is one wavelength. In the same way, the distance between adjacent red circles is one wavelength, and the distance between adjacent red and blue circles is one-half of a wavelength. So, if we know where the crests are, we know where the troughs will be, and vice-versa. We can choose to work with the crests and omit the troughs altogether.

More generally, we can have a simplified picture of the wave by drawing lines connecting points of equal phase, and choose the separation between adjacent lines to be one wavelength. We call these lines the **wave fronts**. We can then visualize the propagation of the wave created in a pond, as a set of circular wave fronts. The wave always propagates in the direction perpendicular to the wave fronts, as indicated by the red arrows in figure 2.1. In the case of the wave in a pond, we have circular wave fronts, with the source at the center. In three dimensions, for instance, the sound from a helicopter, we would have spherical wave fronts, or a **spherical wave**.

Another useful model for wave fronts is the so-called **plane wave**. In figure 2.2, we see that the wave crests are to some extent parallel straight lines. In three dimensions this pattern of wave fronts would correspond to parallel planes. As it is easier to work with parallel lines and planes rather than circles and spheres, these models are very useful because when the wave source is far from the observer, the wave fronts begin to look a lot like plane waves[1]. This fact makes the plane wave a useful model that we will use occasionally (see figures 2.3, 2.5 and 2.6).

Figure 2.2. Pattern of straight parallel wave crests.

---

[1] In the same way, our horizon looks flat, because we are very far from the center of the Earth!

2.3 Sound propagating in air

The most common case of sound propagation is sound propagating in the atmosphere. As mentioned in section 1.2, sound waves in air are a succession of travelling compressions and rarefactions, meaning the total air pressure along the path of the sound wave goes higher or lower than the atmospheric pressure. We will use the term **acoustic pressure** or **sound pressure** (or simply pressure) to indicate the difference between total pressure and atmospheric or ambient pressure[2]. So, at any point along the wave, the total pressure is the sum of the atmospheric pressure plus the acoustic pressure, which is the quantity of interest in our discussion. Air pressure is always a positive quantity. As the total pressure can be higher than the ambient pressure (that happens in a compression part of the wave) or lower than the ambient pressure (that happens during the rarefaction part of the wave), the acoustic pressure can be positive or negative. Compressions or rarefactions can be represented by wave fronts. The speed of the propagating wave fronts is what we call the speed of sound, which depends primarily on the atmospheric temperature. The effect of other factors, such as humidity and atmospheric pressure, is minor.

At 20 °C (about 68 °F) the speed of sound is 343 m s^{-1} (1126 ft s^{-1}). The speed of sound in air increases by about 6 m s^{-1} for every 10 °C increase in temperature. This is equivalent to an increase of 10 ft s^{-1} for a temperature increase of 10 °F. From section 1.1 we know that for a pure sinusoidal waveform, the speed is equal to the product (frequency) × (wavelength). If the air temperature along the path of the wave increases, the speed will increase and the wavelength will increase proportionally, but the *frequency will remain the same.*

Note that while the compressions and rarefactions travel in air at the speed of sound, the molecules in the air do not travel along with the wave. The air molecules do a back-and-forth motion between compressions and rarefactions along the direction of propagation of the wave, and on average the distance travelled by the molecules is zero.

Table 2.1. Speed of sound and density of selected materials.

Material	Speed of sound (m s^{-1})	Density (kg m^{-3})
Air	343	1.2
Water	1500	1000
Cork	400–500	200–300
Wood	3300–3600	500–800
Concrete	3200–3600	2300
Steel	6000	7800

---

[2] Pressure is measured in pascals (Pa.) Normally at sea level the atmospheric pressure is 101 325 Pa.

Sound can also propagate in liquids and in solids. Compared to the speed of sound in air, the speed of sound in liquids is higher, and the speed of sound in solids even higher. Table 2.1 lists some typical values for the speed of sound and the density of the material in kilograms per cubic meter[3]. We note that the speed of sound is higher for the denser and stiffer materials, such as concrete and steel.

2.4 Reflection

2.4.1 Reflection, diffuse reflection and echo

From everyday experience we know that different fractions of the incident light are reflected from different surfaces. A mirror reflects almost all the incident light while a clean glass window will reflect a small fraction (about 4%) of incident light. If the reflecting surface is smooth, the light is reflected in a specific direction, determined by the direction of the incident ray. This is the case of **specular reflection**, or simply **reflection**. The reflection of sound waves is very similar.

We also know that rough surfaces (for example, scratches, or fingerprints, or residue on reading glasses) appear whitish. This is the result of reflections from small irregularities or small particles on the surface. These reflections occur in all directions, and the phenomenon is known as **diffuse reflection** or **scattering**. The same applies to sound waves. In the case of sound waves a textured surface, such as the surface of acoustic tiles, will cause some reflection or scattering, but may cause significant absorption as well.

A familiar example of sound reflection is **echo**, named after a nymph in Greek mythology (shown in the chapter opening figure) who was cursed to *repeat* the last word she heard. We can hear our echo if we stand a distance of at least 17 m away from a wall of a large building. What we hear is the reflection of our voice. Except for special effects, echo is generally an undesirable effect in music performances. If the distance to the wall is less than 17 m, then our hearing is unable to distinguish the reflection as a separate sound. In this case, the reflected sound fuses with the original sound, and the result is that the sound appears to persist for a longer time. This effect is called **reverberation**, and is noticeable in large empty rooms. Reverberation plays a significant role in the quality of acoustics in large halls, and will be further discussed in section 9.4. Echo will occur only if the reflecting surface is large compared to the wavelength of sound. We do not hear our echo from smaller reflecting surfaces; street signs, for example. The interaction of sound waves with smaller objects is discussed below.

In terms of the direction of the reflected wave, sound waves follow the law of reflection, which tells us that the angle of reflection is equal to the angle of incidence of the wave, as illustrated schematically in figure 2.3. The angles are measured from the *normal* to the wall surface at the point of incidence.

---

[3] Density is the amount of mass per unit volume of material.

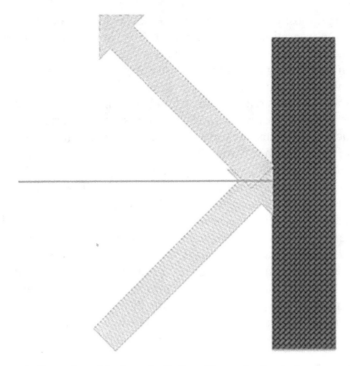

Figure 2.3. Schematic illustration of the law of reflection. The angle of reflection is equal to the angle of incidence of the wave.

2.4.2 Specific acoustic impedance

The law of reflection allows us to determine the direction of the reflected wave, but does not tell us about the amount of energy that is reflected. When a sound wave reaches the boundary of two media, for example between air and the wall in figure 2.3, some of the wave energy will be reflected back to the air, and some of the wave energy will cross the boundary, and propagate into the wall. The fraction of the energy reflected depends on the properties of the two media. The property of interest here is the so-called **specific acoustic impedance of the material**. The impedance of a medium is the product:

$$\text{specific acoustic impedance} = (\text{density of medium}) \times$$

$$(\text{speed of sound in the medium})$$

The specific acoustic impedance tells us how much 'resistance' is encountered by the wave travelling in the medium. For example, using values from table 2.1, we see that the specific acoustic impedance of air is $343 \times 1.2 = 412 \text{ Pa s m}^{-1}$. Similarly, for water the specific acoustic impedance is $1500 \times 1000 = 1\,500\,000 \text{ Pa s m}^{-1}$, which is much larger than that of air. In other words, it is more difficult for sound to travel through water than air.

The percentage of the reflected energy at the boundary of two materials depends on the difference between the impedances of the two materials. If the impedance difference is large, most of the energy will be reflected at the boundary between the two materials. On inspection, we see from table 2.1 that the impedance of air is much smaller than the impedance of any of the materials listed in the table. This means that sound waves in air reaching any boundary will reflect back almost entirely, and only a small fraction will cross the boundary. For example, at the air–water boundary, over 99.9% of the wave energy is reflected back, and less than one part in a thousand gets in the water. The percentage of sound reflected from solid surfaces is even larger. With such high fraction of sound reflected, one may wonder how we can still hear conversations from an adjacent room? The answer is that our hearing can pick up very weak sounds. As we will see in chapter 4, if we reduce the sound level of a normal conversation by one thousand times, the result would still be audible as a whisper. Actually, the situation is more complicated when sound waves travelling in air reflect off a solid surface, because in this case absorption occurs at the surface of the solid. The fraction of energy absorbed depends on the material and the texture of the surface. Generally, softer materials with textured surface such as upholstery absorb significantly more than rigid smooth surfaces, such as a hardwood floor. Usually, the higher frequencies are absorbed more than low frequencies, so when we hear a conversation from an adjacent room, the sound is significantly distorted, and the voices sound 'dull.' Absorption from surfaces is very important in determining the acoustic qualities of a music hall or an auditorium, as will be discussed in chapter 9.

2.4.3 Acoustic impedance and impedance matching

The specific acoustic impedance is a property of the material. The electrical analogue would be the **resistivity** of a material. For musical instruments, such as flutes, trumpets, and the like, there is actually no change in medium. The input is air, and the output is air. What is important in musical instruments is how much 'resistance' is caused by the *shape* of the instrument itself. This is called the **acoustic impedance**. So, what determines the impedance is the shape of the various parts of the instrument. Here again there is analogy to wires. If the wire is thick and short, the electric current flows easily. The same is true for wind instruments. If the bore of a flute is wide, the impedance is low. One may be tempted to think that very low impedance would be better, and might make the instrument easier to play, because there would be no 'resistance.' This is not the case. In fact, a high impedance is necessary for the instrument to make a sound at all! For example, suppose that we are trying to create a sound by blowing across one end of an open pipe. As we will see in chapter 6, sound will build up in a pipe only if part of the sound wave is reflected back into the tube, otherwise, any pressure introduced into the pipe would leak straight out at the other end. Of course, we do not want the entire sound wave to reflect back into the tube either. In short, we need to have the right amount of reflection at the open end.

In chapter 6 we will discuss the tone frequencies that can be generated by pipes. In this section we will focus on what affects the percentage of sound energy that we can get out of a pipe. This percentage depends on how much energy is reflected back into the pipe, in other words, it is related to the *change* of impedance at the open end of the pipe. For a long cylindrical pipe of cross-sectional area A, the impedance inside the pipe is in inverse relation to the cross-sectional area of the pipe. If the cross-sectional area of the pipe changes, for example, if the pipe becomes narrower at some point, then there will be a change in impedance at that point, and sound waves traveling inside the pipe will be partially reflected back at the constriction. In wind instruments we may also have holes on the side of the pipe. This makes the situation a bit more complicated, but the basic mechanism is the same, that is, changes in the acoustic impedance at any point along the path of the sound wave will cause part of the wave energy to reflect back. The strength of the reflection depends on the percentage change of the impedance. An interesting fact is that if the change of the impedance happens in steps, then the reflection is weaker than when the change of impedance happens all at once at one point of the pipe. For instance, suppose that we have a situation where the cross-sectional area changes from 1 to 2. If we want to have less reflection, we can make the change in two steps at two different points of the pipe, for instance, one step could be from 1 to 1.5 followed by a second step from 1.5 to 2. Here we replaced a large jump (1–2) by two smaller jumps, 1 to 1.5 and 1.5 to 2. In a way this smooths out the transition from high to low impedance, and weakens the reflection. We could add more steps in between and make the reflection even weaker.

A major change in impedance happens at the open end of the pipe. One may wonder about the cross-sectional area seen by the wave once it leaves the open end of the pipe, and is no longer confined within the walls of the pipe. Is the cross-sectional area at the end of the pipe 'infinite'? That is not so, because according to our equation above, an infinite cross section would make the impedance zero, and that would reflect all the sound back into the pipe. The answer is (see section 2.6) that the wave spreads out gradually into a cone. In other words, at the open end of the pipe, the wave sees a gradual increase in diameter, like going through a cone.

In some instruments the open end flares out in a cone of sorts. The role of the flare in these instruments (clarinet, trumpet, tuba, and other) is to make the impedance change more gradual. We can imagine a flare as a continuous sequence of small steps, and as mentioned above the result is a smoother transition with less reflection and more output compared to the instruments without flare, like the flute. This gradual change is referred to as **impedance matching**. A very similar reflection of energy occurs also in electrical circuits, and as we will see in chapter 7, it is a key parameter when connecting different sound equipment. For example, when connecting a microphone to an amplifier, we need most of the signal from the microphone to flow into the amplifier, and minimize the reflection of signal back to the microphone. If the two devices have very different impedances, we need to add impedance 'steps' between microphone and amplifier to make the impedance change from the microphone to the amplifier more gradual.

2.5 Interference

As discussed in section 1.3, the result of adding two sinusoids of the same amplitude and frequency depends on the phase difference between the sinusoids (see chapter 1, figures 1.4–1.6). We showed that the amplitude doubles if the sinusoids are in-step.

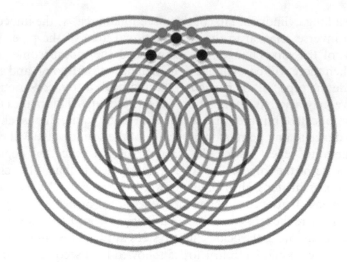

Figure 2.4. Interference between waves produced by two sources. Red and black dots mark points of constructive interference. The blue dots mark points of destructive interference.

If they are 1/2 of a cycle out-of-step, cancellation occurs. For intermediate values of the phase difference, the amplitude of the combined wave is less than the sum of the amplitudes of the individual waves. The results can be directly applied to the situation when two sound waves of the same frequency run into each other at some point. The interaction between two waves of the same frequency is called **interference**.

Figure 2.4 is a snapshot of two waves produced by two sources that are side-by-side. We assume that the two waves have the same amplitude and the same frequency (same wavelength). The gray circles in figure 2.4 indicate the pressure highs (**compressions**) and the blue circles indicate the pressure lows (**rarefactions**) of the two waves. Because the waves have the same wavelength, the distance between successive highs (which is one wavelength) is the same for both waves. The same is true for the blue circles. It follows then that at the intersections of two gray circles the waves are in step, therefore the pressure amplitude is the sum of the two amplitudes at these intersection points. These are points of **constructive** interference. The red dots in the figure mark some points of constructive interference. At the intersections of two blue circles the waves are also in step, and the amplitude is the sum of the two amplitudes at these intersection points. The black dots mark some of these points. At the intersections of a gray and a blue circle, the two waves have phase difference of 1/2 cycle and the amplitude is the difference between the two amplitudes at these intersection points, so the two waves cancel each other to some degree. These are points of **destructive** interference, and are marked by the blue dots in the figure. At all other points in the region where the two waves overlap, the amplitude is less than the sum, but larger than the difference between the two amplitudes, so we have partial cancelation.

The pattern of constructive and destructive interference points shown in figure 2.4 applies to a given wavelength, which is what determines the spacing between the wave fronts. Similar patterns can be obtained for different wavelengths, and the points of constructive/destructive interference would be shifted accordingly. For longer wavelengths (lower frequencies) the spacing of the wave fronts will be wider and the distance between adjacent points of constructive/destructive interference will be larger. The reverse is true for shorter wavelengths (higher frequencies).

Figure 2.4 shows only a snapshot in time for a given wavelength. Because the phase difference between the two waves is determined by the *distance* of a given point from the two sources, the pattern of constructive/destructive interference *does not change* with time. The results of interference can be noticeable in concerts, particularly with the low frequency (that is long wavelength) tones. Listeners may hear too much of a bass tone if they happen to sit at a point of constructive interference for that particular wavelength, and may miss some tones altogether if they happen to sit at points of destructive interference. Note also that destructive interference does not occur along the line midway between the two sound sources. This means that seats midway between the speakers would be the best choice, because at these seats the effects of interference are minimized.

2.6 Diffraction

A shadow occurs when an obstacle blocks the path of light rays. The result is more or less a two-dimensional replica of the object. With proper lighting, the shadows of ordinary objects can appear very 'crisp' leading one to think that light rays do not 'bend' around corners. Actually, all waves, including light and sound waves, can bend around corners[4]. This explains why we can hear sounds around the corners of buildings, even if we don't have direct view of the sound source. This property of waves is called **diffraction**. If a *small* object happens to be in the path of the wave, like a hand in front of one's mouth, the wave will bend around it. In other words, a small object can become 'invisible' to the wave. The measure of what is *small* and what is *large* is the wavelength of the wave we are dealing with. In the case of light, *small* means an object that is hardly visible to the human eye, like a tiny particle of smoke. In the case of sound, *small* can mean something like the size of a large dog.

In terms of music, a very interesting situation occurs when a sound wave goes through an opening. For instance, when a sound wave finds a barrier with a small opening, like a window in a wall. As the wave goes through the opening, it bends around the edges and spreads out. The amount of spread depends on the size of the opening. This is shown in figure 2.5. In both cases the size of the opening is comparable to the wavelength. In figure 2.5(a) the opening is larger compared to figure 2.5(b). The important point here is that the wave spreads over a wider angle when going through the small opening. This is the reason why we cup our hands in front of our mouth; for example, when coaches are giving instructions from the sideline. Our hands create a larger opening than the mouth, and the sound waves (and energy) spread less, that is, the sound is more directional. Again, what is small and large depends on the wavelength. That means that for a given opening, the

---

[4] F M Grimaldi was the first to make this remarkable observation about light in the mid-17th century.

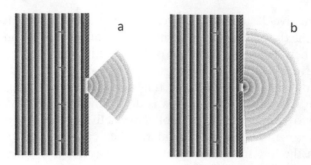

Figure 2.5. (a) Wave diffraction through an opening. (b) The spread is wider for the smaller opening.

longer wavelengths (the lower frequencies) will spread into a wider angle than the shorter wavelengths (higher frequencies). As a direct result, the high frequencies of music from a loudspeaker spread into a smaller angle, so they are louder directly in front of the speaker. The bass notes spread into a wider angle.

2.7 Absorption and attenuation

From everyday experience, we know that a 60 W light bulb may appear dazzlingly bright if it is very close, but appears very dim if it is 1 km away. Whether close or far, the 60 W rating is a property of the light bulb. It describes the rate at which energy is emitted from the light bulb, and is independent of the distance. The energy is emitted by the source at a constant rate, but as energy propagates away from the source, it spreads out in all directions. At 1 m away from the light source, the energy spreads over a sphere of radius 1 m. At a distance of 1 km from the source, the *same* amount of energy has to spread over a sphere of radius 1 km, meaning that a lot less energy crosses each square meter of the sphere at that distance. The power crossing per unit area is what we defined as the intensity (see section 1.5.) It follows that the dependence of the intensity on the distance from the source follows the **inverse square law**, which tells us that the intensity depends on the inverse of the distance squared.

The inverse square law applies to sound waves as well[5]. According to the inverse square law, if the distance doubles, the intensity is reduced to $1/2^2 = 1/4$. If the distance is 10 times larger, the intensity is reduced to $1/10^2 = 1/100$. Table 2.2 lists the intensity for selected distances from a small 10 W sound source[6].

From the values in table 2.2 we see that because of the inverse square law, the intensity decreases rather rapidly. In addition to the loss of intensity due to spreading, the atmosphere absorbs part of the sound energy. The absorption affects primarily the higher frequencies over 1 kHz and increases with frequency. For example, at a distance of 10 m from the sound source, the loss to absorption is 2% for frequency 1000 Hz, 20% for frequency 5000 Hz, and 50% for frequency 10 000 Hz. As the loss due to spreading affects all frequencies more or less the same, distant

---

[5] A sample calculation of the intensity at various distances is given in section 2.9.1.
[6] Here we neglect absorption and other losses for simplicity.

Table 2.2. Intensity from a small sound source of 10 W at selected distances from the source.

Distance from source (m)	Intensity (W m^{-2})
1	0.8
10	0.008
100	0.000 08
1000	0.000 000 8

sounds tend to lose the higher frequencies and sound dull. For example, a listener sitting 30 m (about 100 ft) away from the stage will lose about 50% of the 5000 Hz tones due to absorption alone.

The combined effect of all factors that reduce the intensity (including the inverse square law) is called **attenuation**. The attenuation of a wave going from point A to point B is expressed by the ratio of the intensities at points A and B. As the intensity can vary by factors of millions and billions, it is more convenient to use units that will keep the numbers small. The attenuation is usually listed in units called **decibels**, which will be discussed in section 4.4.

The intensity is an objective quantity, in other words it has a value that can be measured, and that value is independent of the person making the measurement. So, if we are at a certain distance from a loudspeaker and we measure the intensity at that point, and then we double the power output of the speaker, the measured intensity should double as well. What is interesting here is that our perception of the 'volume' change will not double. Also, if two people are there, each will probably perceive a different change in volume. The reason is that our perception of 'volume' is a subjective quantity, and does not have a one-to-one relation to the intensity. This relation will be explored further in chapter 4.

2.8 Further discussion

2.8.1 Refraction of sound waves

As we learned in section 2.3, the speed of sound in the atmosphere depends primarily on the temperature. The temperature of the atmosphere normally decreases as the altitude above ground increases, in which case the speed of sound decreases as well. This means that the lower parts of the wave fronts travel faster than the upper parts, as shown in figure 2.6(a). The result is that the direction of propagation of the wave bends upward. Under certain conditions the lower part of the atmosphere is colder than the upper parts. This temperature behavior is called *atmospheric inversion*, and is common over large masses of water such as lakes. In this case, the lower parts of the wave fronts travel slower than the upper parts, and the direction of propagation turns downward, as if the sound is reflected down to ground level, as shown in figure 2.6(b). In either case, this bending of the direction of propagation is called **refraction**, and is of major importance in the case of light waves. For musical sound indoors the phenomenon of refraction is of little practical importance. It is important for outdoors acoustics, for example in concerts, where the sound can

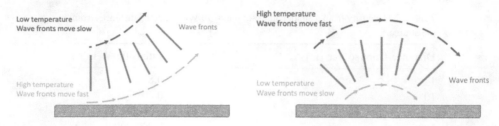

Figure 2.6. (a) Path of sound waves when air temperature decreases with altitude. (b) Path of sound waves in conditions of atmospheric inversion.

bend downward and disturb neighbors, but also bring the noise from nearby highways to the concert venue.

2.8.2 Effects of temperature on wind instruments

As we shall see in sections 6.6 and 6.7, the basic principle in wind instruments is that the tones produced have a wavelength determined by the length of the oscillating column of air inside the instrument. Of course, as the temperature changes the length of the instrument changes, but this change is insignificant. So, the wavelength of the tones produced will not change with temperature. What is significant is that the speed of sound in the instrument changes with temperature. If the speed changes, while the wavelength remains the same, the basic relation:

$$\text{speed} = (\text{frequency}) \times (\text{wavelength})$$

tells us that the frequency *must* change. For instance, a flute in a room where the temperature is 25 °C (77 °F) can produce a frequency of 330 Hz, which corresponds to the musical tone E4. If the temperature in the room is increased to 35 °C (95 °F) the speed of sound will increase from 346 m s^{-1} to 352 m s^{-1}, that is, an increase of 6 m s^{-1} or 1.7%. As the wavelength remains the same, the frequency will increase proportionally, that is, it will increase by 1.7%. In other words, the tone will now have a frequency of 335 Hz instead of 330 Hz, which is quite perceptible. The change in temperature may not affect other instruments, for example, an accompanying guitar. So, we have the flute playing 335 Hz, and the guitar playing 330 Hz for the same note, E4. This makes the combination of the two instruments sound 'out of tune', which is unpleasant to the ear.

2.8.3 Baffles

As a result of diffraction, sound waves can bend around the opening from where the sound is emerging. For example, the sound emitted from the back of a drumhead can bend around and propagate upward. Similarly, for an un-mounted loudspeaker, the sound emitted from the back of the speaker will bend around and propagate forward. In both cases we have a vibrating surface that emits sound in the forward and backward directions. As the surface is vibrating forward it creates compression in front of the surface and rarefaction in the back of the surface. This means that the waves at the front and the back of the surface are 1/2 cycle out of phase. Because the wave from the back bends around and gets to the front side, we will have destructive interference, and the resulting sound will be weakened. This is why drums and loudspeakers, especially the ones used for low frequency (bass) sound must have a barrier, or **baffle**, to reduce the effect of destructive interference.

2.8.4 Scientific models

In section 2.2 different wavefront patterns were introduced, such as the plane waves, circular waves, and so on. We used these concepts to describe the propagation of waves. These are idealized concepts, or **models**, that simplify the mathematics needed to answer specific questions. This is a common practice in science. For example, to understand some phenomena, astronomers usually assume that the Earth is a sphere, which is actually not true. The Earth is a flattened sphere. But this does not matter really if we are trying to predict the phases of the Moon. Similar is the case with the models we introduced to describe the propagation of waves. For instance, plane waves must be infinite in extent, otherwise they would curve by diffraction. Of course, no wave extends infinitely. In the same way both circular and spherical waves are idealized models, because to produce perfect circles or spheres, the source at the center must be a point, that is, it must have zero size. Of course, an object of zero size cannot emit sound or anything else for that matter. Another example of particular interest to string instruments is the description of a vibrating string. To solve the problem of the vibrating string, we start by assuming that the string is perfectly flexible, and also that the amplitude of the vibration is very small. Under these assumptions we have the so-called **linear** response, meaning that the restoring force is directly proportional to the extension of the string. In reality, all strings, especially the thicker ones, have some stiffness, which acts as an additional restoring force. Just the same, as we will see in chapter 6, our model of perfectly flexible strings gives results that are in most cases remarkably accurate. On the other hand, for the thicker strings, for example in a piano or a bass guitar, the effects of stiffness are perceptible by the human ear. In this case one must include the stiffness, and the equation of motion is **non-linear**, which is much more complicated. The effect of non-linear effects in strings will be discussed further in chapter 6. The point here is that simple models may not be exactly correct, but at least they give us a good start to begin understanding mathematically more complex problems.

2.8.5 Sonography

The time it takes for a wave to reflect from an object and return to the source gives us a way of measuring the distance of the reflecting object. The distance equals the speed of sound times one half of the round-trip time. Methods of medical imaging use this 'echo' technique with very high frequency sound waves not audible to the human ear, the so-called 'ultrasounds' to image tissues, blood vessels, organs, and so on. This method is called **sonography**, and relies on the fact there is a change in the density of different parts of the body. A change in density means change in impedance, and as we discussed in section 2.4.2, this means reflection. So, once the wave hits the surface separating, say, a muscle tissue from a bone, the acoustic impedance changes, and the result is a small reflection or 'echo' of the sound wave. This will happen everywhere there is a change in impedance, and the strength of the reflection as mentioned above, is determined by the change in the impedance. By scanning a large area and converting the 'echo' times to distances, one can get a three-dimensional image of the body's interior.

2.9 Equations

2.9.1 The inverse square law

The mathematical equation for the inverse square law is:

$$I = \frac{P_{source}}{4\pi d^2}$$

where P_{source} is the power emitted by the sound source, d is the distance from the source, and I is the intensity at a distance d away from the source.

Example: If $P_{source} = 10$ W and $d = 1$ m then $I = 0.8$ W m^{-2}
   If $d = 10$ m then $I = 10/(4\pi 10^2) = 0.008$ W m^{-2}
   If $d = 100$ m then $I = 10/(4\pi 100^2) = 0.000\,08$ W m^{-2}

The *average* intensity of a sound wave in air is given by the equation:

$$I = p^2/2\rho c$$

In the above equation p is the *amplitude* of the acoustic pressure, c is the speed of sound and ρ is the density of air.

2.9.2 The rms value

We could relate the average intensity to some kind of average of the acoustic pressure. The average value for *any* sinusoidal wave over one cycle is zero, because the positive values of the sinusoid cancel the negative values, so the average value is not useful. The way to go around this problem is to first square the sinusoid, in which case the value is always positive. We can then average the square over one cycle, and then take the square root. The result is called the **root mean square**, or **rms** value for short. As it turns out, the amplitude and the rms value are related by the simple relation:

amplitude $= \sqrt{2}$ rms or amplitude $= 1.41$ rms and rms $= 0.707$ amplitude

We can rewrite the above equations in terms or the rms value of the acoustic pressure as:

$$I = p^2/\rho c$$
$$p = \sqrt{\rho c I}$$

where p now stands for the rms value of the acoustic pressure. In what follows we will use the symbol p to indicate the *rms* values for the acoustic pressure.

2.10 Problems and questions

1. Suppose that the wave represented in figure 2.1 has wavelength equal to 3 meters.
    (a) What is the distance between successive blue circles?
    (b) What is the distance between successive red circles?
    (c) What is the distance between adjacent blue and red circles?
2. Do you expect the echo-time to be longer in a hot desert or in Antarctica? Explain your reasoning.
3. Explain how can we overhear a conversation through an open window, even if we cannot see the people speaking. Would it matter if the people speaking were facing the window or not?
4.   (a) How long does it take for sound to travel a distance of 3.5 km?
    (b) Light travels about a million times faster. How long does it take for light to travel the same distance?
    (c) Suppose we see lightning strike and hear the thunder sound 5 seconds later. How far away was the lightning?
5. Sonography (section 2.8.5) uses sound waves of very short wavelength (ultrasounds) to image very small parts of the body. Why is it necessary to use short wavelengths?
6. In hurricanes the winds can reach speeds over 250 kilometers per second.
    (a) Express this speed in units of m s^{-1}
    (b) How does the air speed in hurricanes compare to the speed of sound?
    (c) Does your result support the assumption that in sound waves the air molecules move along with the compressions and rarefactions of the sound wave?

IOP Publishing

The Physics of Sound Waves (Second Edition)
Music, instruments, and sound equipment
Panos Photinos

Chapter 3

Displaying and analyzing musical sounds

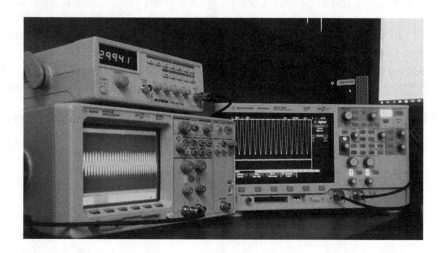

3.1 Introduction

When describing peoples' voices or sounds of instruments, we often use adjectives such as 'warm', 'crisp', 'crystalline', and the like. These words may have an intuitive clear meaning, yet it is hard to quantify these characteristics. In other words, these statements are subjective. A trained ear can often distinguish the sound produced by two seemingly identical musical instruments, and in the not-so-distant past, instrument makers relied on sound perception alone. Today we can measure, analyze, and identify the quantities that make a voice or the sound of a musical instrument distinguishable. Analyzing a sound has many practical applications, such as voice recognition, electronic synthesizers, sound recording and reproduction, and other. This analysis requires some way of measuring and displaying the musical sound.

As we will see, even the simplest sounds produced by a musical instrument can appear very complicated, as illustrated by figure 3.1 which captures the sound of a single note played on a guitar. The note is C4 which corresponds to a frequency of 262 Hz (see appendix C) and lasts about 0.5 s. This graph does not look at all like the sinusoids we discussed so far. For instance, the peaks do not have the same height. Overall, the height increases at first then dies away. We also note some 'wiggles' at seemingly irregular intervals. Are these details of any significance? One might guess the answer by looking at figures 1.7 or 1.8 which show the result of adding two waves of different frequencies. The sum of the two waves shows wiggles, so we might guess that the wiggles we see in figure 3.1 are the result of having more than one frequency present. The number of frequencies present, and other details, such as the overall behavior of the peak height, may contain the key information about this sound that makes it different than the sound produced by striking the same note on a piano or a xylophone. In this chapter, we will consider three basic ways of representing musical sounds, and ways to extract the measurable quantities that give a tone its characteristic quality or **timbre**.

3.2 Measuring sound signals

Sound signals are measured using microphones (see section 7.4) and are usually expressed in units of volts (V). For example, in figure 3.1 the horizontal axis represents time and the vertical axis represents the voltage. We could also convert the voltage into pressure, since sound waves are essentially pressure waves. This conversion would require *calibration*, that is, a way of converting the voltage values into pressure values. Here we will refer to the quantity on the vertical axis as 'voltage' keeping in mind that the important thing is the qualitative variation of what the vertical axis represents without worrying too much about the specific values.

In figure 3.1 the tone corresponds to a nominal frequency of 262 Hz, which is the **fundamental** frequency of the C4 note. Obviously, the graph is not a sinusoid, in other words it is not a **pure tone** consisting of just one frequency. We know that 262 Hz means 262 cycles per second. As the graph captures about 0.5 s in time, we expect to have about 131 positive and 131 negative peaks, which is 262 peaks in total. An immediate question one can have is: how many data points do we need to make this graph? We could start by graphing the peak values, which amounts to 262 points.

Figure 3.1. Sound signal of note C4 played on a *Gibson* acoustic guitar. Duration 0.5 s.

By doing so we may lose the information contained in the 'wiggles' and, as mentioned above, this information is important. As discussed below, the 'wiggles' are caused by the presence of additional components of higher frequency than the fundamental (262 Hz in this example). These are the **overtones**, and have frequencies that may or may not relate directly to the frequency of the fundamental.

More voltage measurements will be required to completely capture the frequency content of this sound. For example, if an overtone has double the frequency of the fundamental, that is 524 Hz, we would need 262 positive and 262 negative peaks for the time interval of 0.5 s captured in the graph. This requires 524 data points in 0.5 s. The conclusion here is that the number of points required to represent the measured tone depends on the frequencies present in a given tone. The problem is that the frequency components of the tone can be determined only *after* the sound has been measured and analyzed. Fortunately, we know that the range of audible sound frequencies normally does not exceed 20 000 Hz, which is the highest frequency one can hear. In other words, for a tone duration of 1 s, we will need 20 000 points for the positive peaks and 20 000 points for the negative peaks, that is 40 000 points in total. For a song that lasts 3 min (180 s) we will need 2 (for positive and negative) × 180 × 20 000 = 7.2 million voltage values! This gives us an idea of why music files take up so much computer memory. We will return to this point in section 10.5.

3.3 Visualizing a simple sound signal

To better understand the different ways of displaying a sound wave, we will start by synthesizing a complex tone using two pure tones (two sinusoids). We will then introduce ways of displaying the parameters of the tone we synthesized. Let us consider a pure tone of frequency 20 Hz shown as wave A in figure 3.2.

The amplitude of the wave is not constant over the duration of the sound. Wave A starts with amplitude 4 and at time equal to 0.15 s the amplitude decreases to 3, and again at time equal to 0.3 s the amplitude decreases to 2 where it remains until

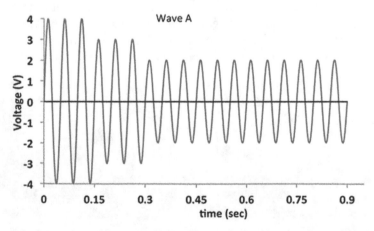

Figure 3.2. A pure tone of frequency 20 Hz. The amplitude changes at time 0.15 and 0.3 s.

the end of the signal. The evolution of the amplitude with time is captured in the three panels of figure 3.3, which show the amplitudes for the first 0.15 s, the amplitude from 0.15 to 0.3 s, and the amplitude after 0.3 s.

The interesting point is that with the information contained in figure 3.3, we can reproduce the tone shown in figure 3.2.[1] All we need to do, is plot a sinusoidal wave of frequency 20 Hz, and use the amplitude values shown in figure 3.3, which are 4 V for the first 0.15 s, then 3 V from 0.15 to 0.3 s, and 2 V from 0.3 s onwards. The meaning of all this is that we have two equivalent ways of representing wave A of figure 3.2. The first way, shown in figure 3.2, is a **waveform view** (or **oscilloscope view**) of wave A. The second way of representing wave A, shown in figure 3.3, is the **amplitude spectrum** of wave A.

At this point we will add a second pure tone of frequency 40 Hz, shown as wave B in figure 3.4. Wave B switches on at time 0.45 s, with amplitude 2. The amplitude increases to 3 at time 0.6 s, and increases again to 4 at time 0.75 s. Figure 3.4 is a waveform view of wave B. The waveform of the complex tone resulting from playing waves A and B simultaneously is shown in figure 3.5.

As wave B switches on at 0.45 s, the combined waveform is identical to figure 3.2 up to time 0.45 s. So, the amplitude spectrum of the waveform is identical to

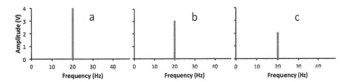

Figure 3.3. Amplitude spectrum of wave A of figure 3.2; (a) for time 0–0.15 s; (b) for time 0.15–0.30 s; and (c) for time 0.30 s and beyond.

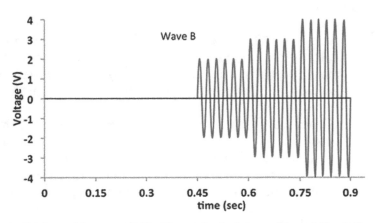

Figure 3.4. A waveform of frequency 40 Hz. The amplitude changes at times 0.45 s, 0.60 s, and 0.75 s.

---

[1] The significance of the phase is discussed section 3.9.1.

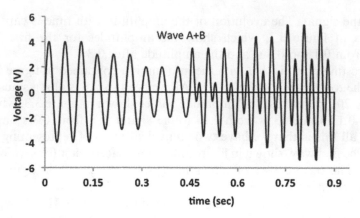

Figure 3.5. Waveform of the complex tone resulting from playing tones A and B simultaneously.

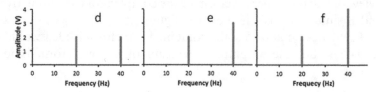

Figure 3.6. Amplitude spectrum of the complex tone of figure 3.5; (d) for time 0.45–0.60 s; (e) for time 0.6–0.75 s; and (f) for time 0.75 s and beyond.

figure 3.3 up to time 0.45 s. After 0.45 s, we note some wiggles in figure 3.5, indicating the presence of the added frequency component from wave B. Figure 3.6 shows the spectrum of the complex tone of figure 3.5 for times after 0.45 s.

We note that the spectrum for times after 0.45 s consists of two lines, one for each frequency component. The blue bars in figure 3.6 represent the amplitude of the 20 Hz tone and the red bars represent the amplitude of the 40 Hz tone. Again, using the information in figures 3.3 and 3.6 one could reconstruct the entire waveform shown in figure 3.5. Obviously, the waveform view is more compact. In our example, we have six amplitude spectra to represent what can be shown by a single waveform view. On the other hand, we cannot readily see the frequencies involved from the waveform view.

Figure 3.7 is a so-called **sonogram** of the waveform of figure 3.5, showing the evolution of the amplitude of each frequency component. Here, the horizontal axis represents time and the vertical axis represents the frequency of each component. In the sonogram we use a color code to indicate the amplitude. Here we will use: orange to indicate amplitude of 4 V, green to indicate amplitude 3 V, and violet to indicate amplitude 2 V.

The waveforms used in our example are not complicated. In real applications, the changes in amplitude may be very small, therefore the color differences may not be as obvious. Figure 3.8 shows a simple sonogram. The sound displayed is the sequence of white keys from C4 to G4 played forward and backward on a

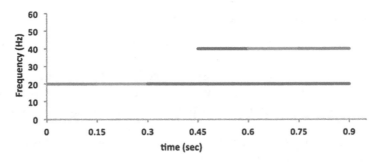

Figure 3.7. Sonogram of the waveform of figure 3.5 showing the evolution of the amplitude of each frequency component. Orange indicates amplitude of 4 V, green indicates 3 V, and violet indicates 2 V.

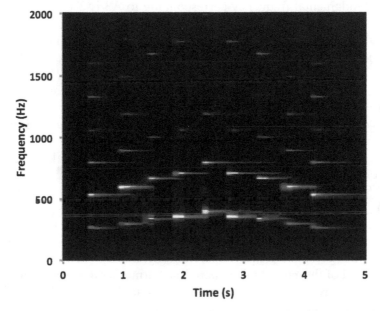

Figure 3.8. Sonogram showing the sequence of white keys from C4 to G4 played forward and backward on a Casio keyboard.

Casio keyboard (the note frequencies are listed in appendix C). So, we have 9 notes in total over an interval of 5 s, indicated on the horizontal axis. The vertical axis indicates frequency. In this diagram we can see anywhere from 3 to 6 frequencies for each note. For each note, the lowest frequency is the fundamental, which is the frequency used in listing the notes in appendix C. Above the fundamental, we have the frequencies of the overtones. The overtones seem to have a regular arrangement in terms of vertical spacing, so one would expect that in this example, there is a simple relation between the frequency of the fundamental and the frequencies of the overtones. The amplitude of each frequency component is color coded, so the resolution is low. The bright yellow indicates the higher values, which typically occur for the fundamental or the first overtone.

3.4 The spectrum of one keyboard note

In many practical situations one usually starts with acquiring the waveform view, that is, measuring the voltage at regular time intervals. In doing so, one must make sure to get enough data points, as discussed in section 3.2. From the waveform view one could find the amplitude spectrum using the so-called **Fourier analysis**, which is a mathematical tool that determines the amplitude of the frequency components of a sound signal[2].

Spreadsheets and more specialized software applications offer this tool under the name FFT (fast Fourier transform) as discussed below. Figure 3.9 shows the amplitude spectrum of the A3 note produced by a Yamaha keyboard set to 'mimic' the sound of a grand piano. The spectrum was taken 0.1 s from the beginning of the signal. The fundamental frequency corresponding to A3 is 220 Hz.

As in the previous figure, there are several frequency components besides the fundamental frequency. The fundamental frequency is 220 Hz, and is indicated by the blue line in the figure. The additional frequency components (the overtones mentioned in section 3.2) are at frequencies 440, 660, 880, 1100, and 1540 Hz, respectively. These frequencies are *integer multiples* of the fundamental frequency, that is $440 = 2 \times 220$ Hz; $660 = 3 \times 220$; $880 = 4 \times 220$; $1100 = 5 \times 220$; and $1540 = 7 \times 220$. As discussed in section 6.11, these frequencies are the **harmonic overtones** or simply **harmonics** of 220 Hz.

Figure 3.10 shows the amplitude spectrum of the A3 note, taken 0.2 s from the beginning of the signal. The amplitudes of all the frequencies are smaller compared to figure 3.9, which represents the spectrum at an earlier instant (0.1 s from the beginning). The result stands to reason, as the entire signal gets weaker with time. Note also that the higher frequencies die out faster than the fundamental. This is especially noticeable for the 880 Hz (cyan line in both figures 3.9 and 3.10) which was the second highest in the earlier spectrum (figure 3.9). We conclude that for this particular sound of the Yamaha keyboard, the harmonics die away quicker than the fundamental. So, as the tone is fading away, it sounds more and more like a pure

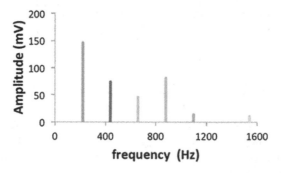

Figure 3.9. Amplitude spectrum of the A3 note produced by a Yamaha keyboard, taken 0.1 s from the beginning of the signal.

---

[2] A brief outline of Fourier analysis is given in section 3.10.1.

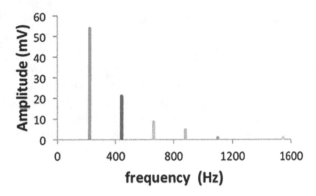

Figure 3.10. Amplitude spectrum of the A3 note produced by a Yamaha keyboard, taken 0.2 s from the beginning of the signal.

tone, that is a tone of a single frequency (220 Hz in our example). This pattern for the harmonics is not the same for all instruments, and this is another feature, beyond the overtone content, that gives an instrument its characteristic sound quality. For example, an A3 note from a guitar, will have different harmonics, with different amplitudes, and different decay patterns for the fundamental and the harmonics, as we will see in the next section.

3.5 Comparing the sound of a steel to a nylon guitar string

As a second example, we will compare the sound of a steel string to a nylon, using the same guitar. We used the bottom string (first string), which when open normally produces an E4, which has fundamental frequency equal to 330 Hz. In both cases the strings are plucked at the same point and the spectra are taken 0.2 s after the beginning of the sound signal.

Figure 3.11 compares the sounds from the two different strings. Figure 3.11(a) shows the amplitude spectrum for the steel string. The spectrum lines show a regular

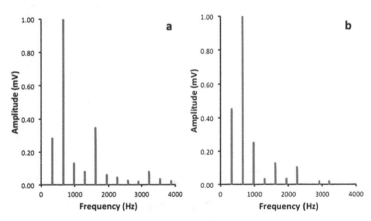

Figure 3.11. (a) Amplitude spectrum of a steel guitar string playing note E4 (330 Hz). (b) Amplitude spectrum of a nylon guitar string playing the same note.

spacing, at integer multiples of the fundamental frequency (that is, 330, 660, 990 Hz, and so on.) We note also that the amplitude of the fundamental is not the largest of all. Instead, it is the first overtone, at frequency 660 Hz, that has the largest amplitude. This is also true of the spectrum of the nylon string, shown in figure 3.11(b). We also note that the amplitude of the fundamental, compared to the higher frequency overtones, is relatively larger for the nylon string. This is what makes the sound quality of the nylon string 'softer'. On the other hand, looking at the higher overtones, we note that the steel string has generally more high frequency components, which gives the steel string its characteristic 'bright' or 'metallic' quality.

3.6 Complex tones and timbre

The sinusoidal waveform is a useful model for pure tones. As was discussed in the previous sections, the tones produced by each instrument can be expressed as a combination of pure tones. A tone consisting of many pure tones is a **complex tone**. Figure 3.12 shows how the sound of a note might begin and end. We note that the sound builds up for the first two cycles; it remains steady for the next six cycles, and then dies out. Is this a pure tone? The answer is *no*. As can be verified by looking at the symmetry of the peaks (in the figure the red dashed lines are given as a guide to the eye): the six peaks in the middle are symmetrical, as a sinusoid *must* be. The first and second peaks are skewed to the left, while the last two peaks are skewed to the right. This means that they are not sinusoids, and the same applies to the waveform as a whole. So, the waveform shown in the figure is a complex tone.

We could also try to get an idea of how the amplitudes of a complex tone evolve by drawing a line connecting the high peaks of the voltage graph. This is done in figure 3.13. The red line connecting the highpoints is called the **envelope** of the waveform.

Figure 3.13 shows four stages of the envelope. At the start we have the **attack**, followed by the **decay**, then comes the **sustain** part, and finally the **release**. To some extent, the performer can control some features of the envelope. This is illustrated in

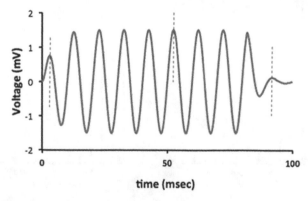

Figure 3.12. A waveform that builds up and dies out. The first and last two cycles are skewed, meaning that the waveform is not a sinusoid.

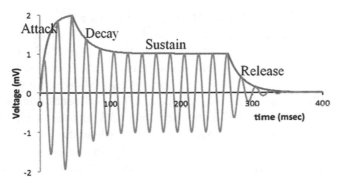

Figure 3.13. The red graph is the envelope of the waveform.

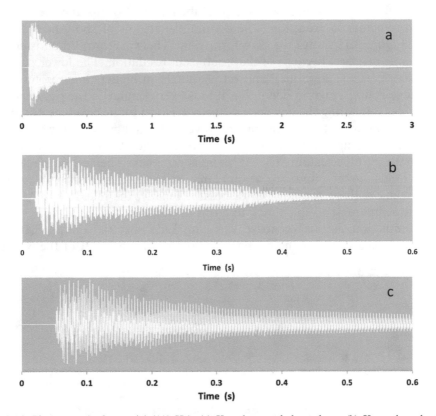

Figure 3.14. Piano sound of note A4 (440 Hz). (a) Key depressed throughout. (b) Key released at once. (c) Early stage of part (a) in higher time resolution looks very similar to (b).

figure 3.14 that shows a keyboard sound of the note A4 (440 Hz). In figure 3.14(a) the key was depressed for the entire recording, and it took about 3 s before the sound died out. In figure 3.14(b) the key was released at once, and the sound lasted a little over 0.5 s. The attack stage is very short, about 0.1 s, and is followed by a brief decay stage, lasting about 0.2 s. The attack in figure 3.14(a) may look very different from figure 3.14(b). Actually, the difference is due to the scale. Figure 3.14(c) shows the attack part of figure 3.14(a) in higher time resolution, which looks very similar to the early part of figure 3.14(b). The real difference between figures 3.14(a) and (b) is in the sustain and release parts. Figure 3.14(a) has a very long sustain stage that declines slowly, while 3.14(b) shows a very brief sustain stage, lasting to about 0.35 s, followed by a release stage that lasts about 0.1 s. The difference comes from the way the note was sounded. As we will see in section 12.3.3, the piano has dampers, that mute the vibration of the string once the key is released, which is exactly what happened in figure 3.14(b). In figure 3.14(a) the key was depressed throughout, and the sound was sustained until the vibrational energy of the string was exhausted.

3.7 The spectrum of a flute

Not all instruments have the sharp attack of the pianos. For instance, let's consider the *quena*, which is a traditional flute from the region of the Andes in South America. Figure 3.15 shows the waveform view. The waveform graph is crowded, and has some artefacts because of the insufficient resolution. The lower part of the figure shows parts of the attack and release stages in higher time resolution. The sustain stage is not shown in detail, but it looks very similar to the right end of the attack stage. Comparing the envelope of this waveform to the envelope of the guitar waveform in figure 3.1 and the piano waveforms in figure 3.14 we see major differences. So, compared to the guitar and piano, the attack stage for the quena leads smoothly to the sustain. That is, there is no decay. The sustain stage is long, actually the sustain for a flute can be very long, as long as one can keep blowing into the instrument without running out of breath. Finally, the release is very distinct, and about comparable to the attack in terms of duration.

For completion, let us compare the amplitude spectra. Figure 3.16 shows the amplitude spectrum for the quena taken 0.2 s after the start of the waveform.

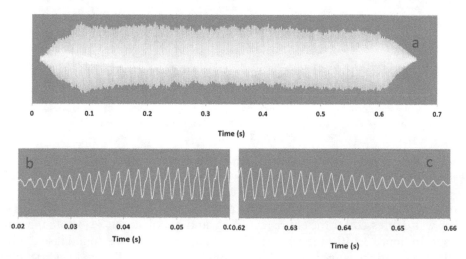

Figure 3.15. (a) Waveform view of a C5 note (523 Hz) from a quena flute. (b) Time resolved attack stage. (c) Time resolved release stage.

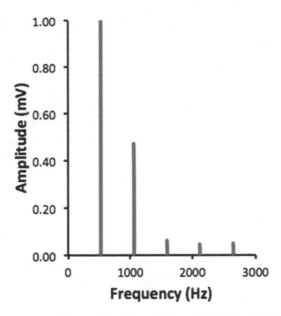

Figure 3.16. Amplitude spectrum of the waveform shown in figure 3.15, taken 0.2 s after the beginning of the sound signal.

The amplitude spectrum shows the lines corresponding to the fundamental frequency and its overtones. Here the frequency of the fundamental is 523 Hz. The lines are regularly spaced, which tells us that the frequencies of the overtones are integer multiples of the fundamental. Note also that in contrast to the guitar spectra of figure 3.11, in the case of the quena flute the fundamental frequency has the largest amplitude. This comparison tells us that the fundamental frequency may or may not be the loudest. In terms of the relative magnitude of amplitudes, the spectrum of the quena comes closer to the piano spectrum of figure 3.10.

As mentioned earlier, some instruments allow the player some control over the envelope, and even the amplitude spectrum (see section 6.4) but there is very little that one can do to make the attack stage of the piano look like the attack stage of a flute, or the sustain stage of a guitar look like the sustain stage of a flute. The point of all this is that the shape of the envelope is a main characteristic of sounds produced by different types of musical instruments, and is a significant tool in synthesizing sound electronically, as will be discussed in section 12.5. So, in conclusion, the quality or **timbre** of a sound played by an instrument is determined not only by the fundamental and overtone frequencies, but also by *how these frequencies build up and die out.*

3.8 Sound analysis software applications

In the laboratory, measuring and analyzing signals, including sound signals, are done using basic instrumentation, that is, an oscilloscope for waveform view (or oscilloscope view), and a spectrum analyzer for the amplitude spectrum (FFT).

Getting the amplitude spectrum from the oscilloscope view can also be done using data analysis software, as mentioned in section 3.4. For sound applications specifically, several apps are available online. A search for 'sound analyzer' will locate a number of apps (including open-source software) suitable for use on computers and laptops, and also for tablets and phones. Most of the apps record user-generated sound using the built-in microphone of the device, and automatically display the *sound* (which is what we referred to as *waveform view*). The signal level may be in volts, millivolts, or decibels (see section 4.4). The spectrum may appear in the menu under FFT, and the sonogram may appear under TFFT.

3.9 Further discussion

3.9.1 The phase of the frequency components

The amplitude spectra tell us about the amplitude of the frequency components of the waveform. In section 1.3, we saw that when adding two waveforms of the *same frequency*, we needed to consider the phase difference between the two waves. Figure 3.17(a) shows the result of adding two waves of different amplitudes and frequencies. The frequency of the blue waveform is 50 Hz and that of the red waveform is 150 Hz. In other words, the higher frequency is an integer multiple of the lower frequency and this is of interest in musical sound. The outcome of combining the two waveforms is shown at the bottom of the figure (green graph). Now suppose that we move the fundamental 1/4 of a cycle back; in other words, change the phase of the blue waveform, keeping everything else the same. The result is shown in figure 3.17(b). The resulting waveform (green graph) has a rather different appearance from the corresponding green graph of figure 3.17(a), although the pattern obviously repeats with the same period of 20 ms. What is interesting here is that for sounds of low intensity the two green waveforms will sound the same! In other words, in general our ear is not sensitive to the phase of the harmonic overtones of the waveform. This means that having the amplitude spectrum is enough information to reproduce the complex tone, and we do not need to worry about the phase difference between the frequency components of the waveform.

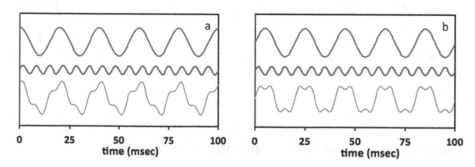

Figure 3.17. (a) Adding two waves (red and blue waveforms) of different amplitudes and frequencies. The green waveform shows the result of the addition. (b) Here, the fundamental (blue waveform) is delayed by one quarter of a cycle.

3.9.2 Playing music backwards and hidden messages

If we play in sequence three keys on a piano, say CEG, we have a pleasing sound of the so-called C the major triad. Also, if we play the sequence in reverse, that is GEC, we have a pleasing sound as well. Could it be that the sequence of notes of a musical masterpiece would produce another masterpiece if played backwards? Maybe! What is clear though is that if we record a sound, even a single key stroke, and play it in reverse, the result has little to do with the original sound. We can understand this fact by looking at figure 3.1. We see that the evolution is not symmetric. Going from left to right, we have the attack, which builds up quickly and finally the sound fades away. Playing the note backwards, that is going from right to left, we see that the sound builds up slowly, then suddenly gets high, then dies away abruptly. The following two audio recordings illustrate the point.

Audio 3.1 plays a C (256 Hz) on a piano, and then plays the sound in reverse.

Audio 3.2 does the same thing for a Ukulele. In both cases, it is very hard to recognize the instrument when the waveform is played in reverse.

One may wonder if the same applies to speech. The next two audio recordings give two different examples. Audio 3.3 is a recording of the word 'mom' played forward and reverse.

Audio 3.1. Piano note C (256 Hz) played forward and reverse. Available at https://iopscience.iop.org/book/978-0-7503-3539-3.

Audio 3.2. Ukulele note C (256 Hz) played forward and reverse. Available at https://iopscience.iop.org/book/978-0-7503-3539-3.

Audio 3.3. The word 'mom' played forward and reverse. Available at https://iopscience.iop.org/book/978-0-7503-3539-3.

Audio 3.4. The word 'cat' played forward and reverse. Available at https://iopscience.iop.org/book/978-0-7503-3539-3.

This combination of sounds seems to be recognizable in reverse. Audio 3.4 does the same for the word 'cat' and this does not work well at all in reverse, particularly for the consonants.

The suggestion that some music lyrics contain hidden messages, from evil sources, has been made on occasion. From the sound of it, one could say that both music and lyrics sounded in reverse are certainly lacking in terms of quality, and *infernal* may be a suitable adjective.

3.10 Equations

3.10.1 Fourier analysis

The basic idea in the analysis of sound is that any periodic function can be expressed as a sum of sinusoids. The frequencies of the sinusoids are all integer multiples of the fundamental, which is the frequency of the periodic function under consideration. This statement is known as the Fourier theorem. The examples of waveforms in this chapter are not periodic functions, because they have a beginning and an end, and the peak levels are not constant. If we take a 'slice' of the signal at any point in time, for example, the part of the signal contained between two successive peaks, we can create a periodic function by repeating the slice indefinitely in time. The period of this function is the time interval between the two peaks selected, and the fundamental frequency is the inverse of the period. If the time interval between the two peaks defining the slice is 0.1 s, then the fundamental frequency is 1/0.1 = 10 Hz. So, according to the Fourier theorem, the waveform for that particular time (say, the center of the slice) can be expressed by a sum of sinusoids, like[3]

$$A \sin(2\pi 10t) + B \sin(2\pi 20t) + C \sin(2\pi 30t) + D \sin(2\pi 40t) + \ldots$$

here the remaining terms are *sine* functions with frequencies 50, 60, 70 Hz, and so on. The sum may also contain *cosine* terms following the same frequency pattern as above. The frequencies are integer multiples of the fundamental. The Fourier theorem also prescribes the procedure to calculate the amplitudes A, B, C, and so on, in the sum. We will first consider the case where the amplitudes are known.

Example: We will revisit the waveform of section 3.3, where we mixed two signals of frequencies 20 and 40 Hz, as shown in figure 3.5. If we use the interval between the last two high peaks as our slice, then our Fourier sum describing the waveform in that slice of time is

$$2 \sin(2\pi 20t) + 4 \sin(2\pi 40t).$$

If we take the slice defined by the first two high peaks in figure 3.5, the Fourier sum describing the waveform in that slice of time becomes

$$4 \sin(2\pi 20t) + 0 \sin(2\pi 40t).$$

The above two equations describe the signal in the **time domain**. The corresponding amplitude spectra (figure 3.6(f) and figure 3.3(a), respectively) describe the signal in the **frequency domain**. The meaning of all this is that the Fourier theorem is essentially a procedure to convert a signal representation from the time domain (the waveform view) to the frequency domain (the amplitude spectrum), and vice versa.

3.10.2 Calculating the amplitudes

As mentioned above, the Fourier theorem gives us the procedure to calculate the amplitudes. This procedure uses integral calculus, and will not be described here. Practical applications use algorithms, which are essentially a sequence of numerical steps, to process the data points of the waveform. The FFT is an algorithm to find the amplitudes. Compared to the integral approach, the FFT calculates the amplitudes much faster and without sacrificing accuracy. As mentioned in section 3.2, our 'slice' must contain a sufficient number of data points to account for all the frequencies that may be included in the signal. As it turns out, the data must be taken at a rate equal to (or higher than) twice the value of the highest frequency component. For example, if we want to make sure that all the frequencies up to 10 000 Hz are captured, then our slice must contain 20 000 data points per second. This would be

---

[3] A sinusoidal oscillation of frequency f can be described by $\sin(2\pi ft)$ or $\cos(2\pi ft)$.

described as a 'sampling rate of 20 kHz'. More discussion on the minimum number of data points required (or minimum sampling rate) can be found in section 8.10.

3.10.3 The frequency components of a square waveform

We can use the Fourier method to construct any periodic waveform. Here we will illustrate the point by adding sinusoids to get a square waveform. This type of wave is a 'model' because if you look *close enough*, nothing in nature has square corners. In any case, the square wave is useful in approximating signals in digital systems and other applications. As it turns out, the Fourier sum of this square wave contains *sin* terms of odd frequencies. Video 3.1 shows how the approximation to a square wave gets better as we add more sinusoids. In this case, a total of 15 terms are added.

The amplitude spectrum is shown in figure 3.18. Note that the amplitudes decrease for the higher frequencies. So, the highest frequency shown here, which is about 31 times the fundamental, still makes a contribution of about 3%. Note that the waveform we analyze here is symmetrical, in other words, the positive and negative cycles are equal in duration. If this were not the case, for example, if the positive cycle was very brief and the negative cycle longer, then the amplitudes would decrease much slower, in other words, the amplitude of all the harmonics would be larger. This is significant for our purpose. As we will see, in many wind instruments for instance, the air input is in the form of brisk short puffs. And our conclusion here is that these puffs contain a large number of frequencies. We will return to this point in chapters 11 and 12.

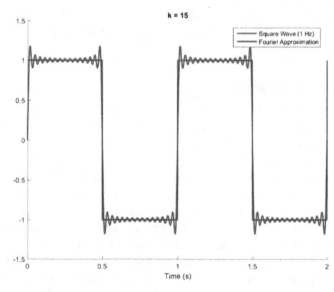

Video 3.1. Fourier sum for square wave of frequency 1 Hz. Video available at https://iopscience.iop.org/book/978-0-7503-3539-3.

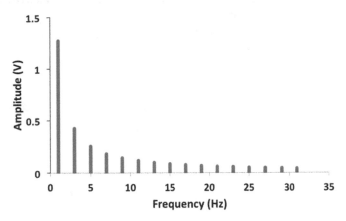

Figure 3.18. Amplitude spectrum of square waveform shown in video 3.1.

3.11 Problems and questions

1. Suppose we want to make a waveform plot of a human voice. We know that the highest frequency contained in that voice is about 5000 Hz. What is the minimum number of points needed to plot the waveform for 1 s?
2. A sound consists of two frequencies, 100 and 110 Hz. The amplitudes are 5 and 2 Volts, respectively.
   (a) Is this a pure tone? Why, or why not?
   (b) Does the waveform change with time?
   (c) What would the amplitude spectrum of this tone look like?
   (d) Does the amplitude spectrum of this tone change with time?
3. Consider the waveforms shown in figure 3.1 (guitar) and 3.15 (flute).
   (a) Which of the two sounds is a pure tone?
   (b) From figures 3.9 and 3.10 we know that the amplitude spectrum of the keyboard changes with time. Do you expect the amplitude spectrum of the flute (figure 3.15) to change to the same degree as time elapses? More? Less? Explain your answer.
4. Examine the waveforms and amplitude spectra discussed in this chapter.
   (a) Which figures represent pure tones?
   (b) Which waveform contains the largest number of frequency components?
5. (a) Examine the waveforms of the guitar (figure 3.1), the piano (figure 3.14), and the flute (figure 3.15.) Which waveform would come closer to sounding the same when played forward and reverse? Explain your answer.
   (b) In audio 3.3 and 3.4 we note that the word 'mom' sounds almost the same when played in reverse, but the word 'cat' does not sound like 'tac' when played in reverse. Use the concept of the envelope to explain why.
   (c) Can you think of words that would not sound the same in reverse?
6. The waveform of the guitar (figure 3.1) starts with high peaks (attack) which gradually diminish. The waveform if the flute (figure 3.15) builds up slowly and has a long sustain stage. Can you explain what causes the difference?

IOP Publishing

The Physics of Sound Waves (Second Edition)
Music, instruments, and sound equipment
Panos Photinos

Chapter 4

The perception of sound

4.1 Introduction

Ordinarily sound comes to **our inner** ear from pressure waves through air, and partially through our bones. Sound waves in air are longitudinal waves consisting of alternating sequences of highs (**compressions**) and lows (**rarefactions**) relative to the atmospheric pressure. Our hearing picks up the pressure pattern, the so-called

doi:10.1088/978-0-7503-3539-3ch4

acoustic pressure (see section 2.3) and through complex processing turns it into what we perceive as sound. As discussed in chapter 1, a sound wave in air is characterized by the frequency content, and the intensity, which can be controlled and changed independently. These are *objective* quantities, and do not depend on the person experiencing them.

In terms of what we perceive, we can distinguish three main characteristics: loudness, pitch, and timbre. These are *subjective* quantities, and depend on the person experiencing them. To understand the difference, we will use the output of a loudspeaker as an example. Suppose that a listener is at some point in the room where the sound intensity is 1 W m^{-2}. If we double the output of the loudspeaker, the intensity at that point will increase to 2 W m^{-2}. The sound intensity is an objectively measurable quantity, and using the proper gadget, the listener could tell that on the scale of Watts per square meter, the reading went from up from 1 to 2. Now if we ask the listener to describe this doubling of intensity, chances are that the listener perceived an increase that is nowhere near doubling, but somewhere in the range between 10% and 50%, say 20%. Of course, we have no device to verify the accuracy of the number the listener is giving us. We just have to take the listener's word for it. If we repeat the experiment with a second listener, the perceived increase would be in the same range, say 40%. Repeating the measurement many times, we might find that on average the answer is 30%. What is interesting here is that if we repeat the experiment with a sound that has a lot of bass, or sound that is barely audible, the results will not be 30%! So, the perception of sound intensity depends on the person, and also depends on *how the measurement is done*, for instance the result may depend on the frequency range used. The conclusion here is that there are no direct and widely applicable relations between the *objective* characteristics of sound and the *subjective* characteristics of the perceived sound. Instead, we have some clearly established trends, and on average, these trends apply to a wide range of circumstances. The scientific study of the relations between subjective and objective attributes of sound is the domain of **psychoacoustics**. Some elementary concepts and results of psychoacoustics will be discussed in the following sections. We will also describe the auditory system, compare it to the visual system, and briefly discuss the use of sound and music and in healing.

4.2 Audible frequencies

The compressions and rarefactions incident on our ear set the eardrum into vibration that follows the pressure pattern of the incident acoustic wave. For a sinusoidal wave, the rate of arrival of compressions and rarefactions is what we defined as frequency in section 1.2. The ear responds to the frequency of the incoming wave, *not* the wavelength of the sound wave. The frequencies we normally perceive range from 20 Hz to 20 kHz[1]. This range is referred to as the **audio or audible frequency range**. In terms of frequency, sounds can be roughly classified as:

---

[1] 1 kHz = 1000 Hz.

Bass	20–250 Hz
Midrange	250–2000 Hz
Treble	2000–20 000 Hz

Sounds of frequency above 20 kHz are referred to as **ultrasounds**, and sound frequencies below 20 Hz are **infrasounds**. Some species can hear well into the ultrasound range; for example, dogs can hear up to 40 kHz, cats up to 70 kHz, and dolphins over 100 kHz. Large animals, like elephants, can also hear infrasounds. The range of audible frequencies varies largely between individuals. With progressing age, the ability to hear high frequencies (above approximately 10 kHz) diminishes.

An ordinary graph of the ranges of the audible frequencies is shown in figure 4.1. Obviously, the low frequency range is very compressed in this graph. To get an idea of how skewed this graph is, let's just say that the blue range in this graphic comprises the first 39 keys (starting from left) of a standard piano keyboard, the green range comprises 36 keys, and the red range only 13 keys.

A better choice of the frequency axis is shown in figure 4.2, where the ranges are about equal in size. In figure 4.2, each step of the frequency axis corresponds to *multiplying* by a factor of 10. This kind of axis has a **logarithmic** scale. In figure 4.1 each step in the scale of the frequency axis represented an *addition* of 2000. This kind of axis has a **linear** scale. When using a logarithmic scale, in essence we are plotting the logarithm of the value, hence the name. The logarithmic scale is very convenient when graphing quantities that cover a very wide range of numerical values, such as the range of audible frequencies in our example.

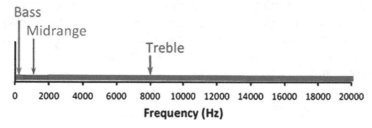

Figure 4.1. Range of audible frequencies using a linear scale for frequency.

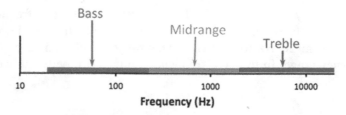

Figure 4.2. Range of audible frequencies using a logarithmic axis.

4.3 Audible intensities

The **intensity** of a sound wave was defined in section 1.5 as the rate of energy flow through an area of one square meter[2]. The intensity is measured in units of Watts per square meter (W m^{-2}). The **threshold of hearing** (TOH) is the lowest sound intensity that can be heard by a human, on average. The TOH depends on the frequency of the sound, as discussed below. This means that for the TOH we need to specify both the intensity and the frequency of the sound. For a sound frequency of 1000 Hz, the TOH is about 2 trillionths of 1 W m^{-2}. In scientific notation, we can write:

$$\textbf{Intensity of TOH} = \textbf{2} \times \textbf{10}^{-12}\,\textbf{W m}^{-2}.$$

According to the inverse-square law (section 2.7) and ignoring absorption, this is the intensity from a 10 W source set at a distance of about 600 km!

The highest intensity that can be experienced without pain or damage is about 10 W m^{-2}. This is the **threshold of pain**. As the name indicates, higher intensities will cause severe pain and potential damage to the ear. This intensity corresponds to a 10 W source at a distance 0.6 m. We see that our hearing works with remarkably *low* intensities. Note also that the threshold of pain is about 10 trillion times larger than the TOH at 1000 Hz, which tells us that our hearing operates over a very *wide range* of intensities. In practice, it is common to use the ratio of the intensity of the sound in question divided by the **reference sound intensity, which is 10^{-12} W m^{-2}**. We will refer to this ratio as *the intensity ratio* or simply *the ratio*.

From the above, it is clear that the range of values of the intensity ratio is huge. So, when making graphs of the intensity ratio, we should consider using a logarithmic scale, as we did with the audible frequency range in figure 4.2, or graph the *logarithm* of the intensity ratio. The latter is the most common choice, because as it turns out, the human perception of measurable quantities, such as the intensity, relates more directly to the logarithm of these quantities, as will be discussed below.

4.4 The decibel (dB) unit

If we take the base-10 logarithm of the intensity ratio of a given sound and multiply it times 10, the result is the intensity of the sound in **decibel (dB)** units[3]. We will use this procedure to determine the dB value for two extreme cases[4].

Example 1: The dB value of the TOH at 1000 Hz.
   The intensity of the TOH is 2×10^{-12} W m^{-2}.
   The reference sound intensity is 10^{-12} W m^{-2}.
   The intensity ratio for the TOH is $2 \times 10^{-12}/10^{-12} = 2$
   The logarithm of the intensity ratio is log2 = 0.301
   So, the TOH corresponds to $10 \times \log 2 = 3.01$ dB.

---

[2] We assume that the area is set perpendicular to the propagation direction of the wave.
[3] A more mathematical discussion on using the dB is given section 4.14.2.
[4] Recall that log1 = 0, log10 = 1, log10^2 = 2, log10^3 = 3, log10^4 = 4 and so on.

Example 2: The dB value of the threshold of pain.
The intensity at the threshold of pain is 10 W m^{-2}.
The reference sound intensity is 10^{-12} W m^{-2}.
The intensity ratio for the threshold of pain is 10/10^{-12} = 10^{13} (ten trillion).
The logarithm of the intensity ratio is log10^{13} = 13
So, the threshold of pain corresponds to 10 × log10^{13} = 10 × 13 = 130 dB

Table 4.1 lists the intensity ratios and dB values for a few examples of sounds.

Table 4.1. Decibel values and intensity ratios for selected sounds.

dB	Intensity ratio	Example
0	1	Light breathing
30	1000 (10^3)	Whisper
60	1 million (10^6)	Normal conversation
90	1 billion (10^9)	Lawn mower at 1 m
120	1 trillion (10^{12})	Jet engine at 10 m

To find the intensity ratio for a sound of given dB value, we reverse the above procedure, that is, we divide the dB value by 10, and use the result as the exponent of 10 (see also section 4.14.2.)

Example 3: Find the intensity ratio of a sound of 60 dB.
First, we divide 60 by 10, 60/10 = 6, and use the result as exponent of 10, that is 10^6.
What this result means, is that the intensity of a 60 dB sound is 10^6 (one million) times the reference sound intensity, in other words, 10^6 × 10^{-12} = 10^{-6} W m^{-2} (one millionth of W m^{-2}).

We can also use the dB values to compare sound intensities. To compare two sounds, we take the difference of their dB values, divide the difference by 10, and use the result as the exponent of 10. The sound with the larger dB has the higher intensity.

Example 4: Compare the intensity of a 90 dB sound to a 30 dB sound.
The difference in dB is 90 − 30 = 60.
We divide the difference by 10, that is 60/10 = 6.
Use the result as exponent of 10, that is 10^6 = 1 000 000.
So, the intensity of the 90 dB sound is 1 000 000 times greater.

As a rule, when comparing two sounds, a difference of:
  10 dB means that the sound intensities differ by a factor of 10.
  20 dB means that the sound intensities differ by a factor of 100.
  30 dB means that the sound intensities differ by a factor of 1000 and so on.

The dB values derived from the intensity ratio represent the **sound intensity level** or **sound power level**. As discussed in section 4.14.1, we can use the **sound pressure level**, (**SPL** for short) instead of the intensity level. In terms of decibels the scale remains essentially the same.

4.5 Frequency dependence of the threshold of hearing

The sound intensity is an objectively measurable quantity. On the other hand, what we generally refer to as 'level' or 'volume' of sound, is the *perception* of the stimulus, which is a subjective matter. For example, two sounds can have the *same intensity* level, yet they may *not* be perceived as being *equally loud*. The numbers listed here are typical values for pure tones, and may change considerably depending on the individual and the method of testing.

Table 4.2 lists representative values for the TOH for selected sound frequencies in the audible range. We note that there is a strong dependence of threshold values on the frequency. A low number of dB indicates that our hearing can pick up low sound intensities (meaning, it is more sensitive) at the corresponding frequency. Our hearing is most sensitive for frequencies in the range of 1000–5000 Hz. The sensitivity is lost for lower frequencies. Note, for example, that at 30 Hz we need at least 60 dB for the sound to be audible at all. As calculated earlier (see example 3) 60 dB means 10^{-6} W m^{-2}, and this is about half a million times the TOH at 1000 Hz (TOH at 100 Hz is 2×10^{-12} W m^{-2}.) This tells us that a 30 Hz sound needs to be at least half a million times more intense than a barely audible 1000 Hz sound to be audible at all! From the values listed in table 4.2, we see that if somehow the sound intensity level of a stereo system is limited to 10 dB, we will not hear any of the frequencies below 250 Hz. In other words, we would *lose the entire bass range*! Above 5000 Hz a similar loss of sensitivity occurs progressively as we approach the upper limit of the audible range. The threshold values listed above are for sound durations longer that 0.5 s. For sound durations shorter than 0.5 s, the threshold values become higher, meaning that our hearing becomes less sensitive for very brief sounds.

Table 4.2. The TOH for selected sound frequencies in the audible range.

Frequency (Hz)	Sound intensity level (dB)
30	60
250	10
1000	3.01
10 000	14

4.6 Loudness level

Note that the above discussion refers to the TOH, and the essential point was that for barely audible pure tones, our hearing responds to much lower intensity levels in the frequency range of 1000–5000 Hz. The question we address in this section is: what happens above the barely audible sounds? One would be correct guessing that the same trend prevails in terms of frequency dependence, in other words, we can expect our hearing to be less sensitive in the lower and higher ends of the range of audible frequencies. To make the discussion more specific, we need a scale that tells us the **loudness level** of the sound, or, how *far above* the TOH the sound is. The values listed in table 4.2 represent the TOH at the corresponding frequencies, so, they all correspond to the *zero-loudness level* at the listed frequency. Next, we need to establish loudness levels, and units, to describe what happens above the zero level. As the perceived loudness level depends on the frequency, it is necessary to adopt a reference frequency before we can construct a loudness level scale. The standard choice of frequency is 1000 Hz. The unit of the loudness level is the **phon**, and is defined in the following way:

A tone of 10 dB (and frequency 1000 Hz) has a loudness level of 10 phon.

A tone of 20 dB (and frequency 1000 Hz) has a loudness level of 20 phon, and so on.

By definition then, for tones of frequency 1000 Hz the intensity level and the loudness level are the same number. This is not true for other frequencies. For example, suppose that we have a 1000 Hz tone of loudness level 20 phon. What should be the intensity level of a 30 Hz tone in order to sound as loud as the 1000 Hz tone? Or, what would be the dB value for the 30 Hz tone in order to have a loudness level of 20 phon? Experiments show[5] that on average, the intensity level of the 30 Hz tone should be about 80 dB for it to sound as loud as the 1000 Hz tone of 20 dB. This result means that the intensity level difference is $80 - 20 = 60$ dB, and from example 4 of section 4.4, we know that a difference of 60 dB, means a factor of 1 000 000. This means that the intensity of the 30 Hz sound should be 1 000 000 times higher for it to sound as loud as the 20 dB, 1000 Hz tone! The results of such measurements are given in a set of graphs, the **equal loudness level contours**, which are discussed in section 4.13.1. For our purposes, we list in table 4.3 representative values for loudness level 40 phon for selected frequencies. We note that tones of frequencies in the high and low end of the audible range must have very high intensity levels (dB) to sound as loud as a 40 dB tone of 1000 Hz.

Table 4.3. Intensity level values (dB) required to produce loudness level 40 phon.

Frequency (Hz)	Loudness level (phon)	Intensity level (dB)
30	40	90
250	40	45
1000	40	40
10 000	40	55

---

[5] Se section 4.13.1.

4.7 Loudness

In the above discussion we are essentially asking how much should the intensity level (dB) be for a tone of a certain frequency in order to have the same loudness level (phon) as a given 1000 Hz sound. We can also ask another important question. Suppose that we have a tone of 1000 Hz, of a certain intensity level, say 40 dB. How much should we increase its intensity level (dB) to make the tone sound twice as loud? What sounds 'twice as loud' is again a subjective matter, and like the loudness level and phon, quantifying **loudness** is based on averaging the results of numerous tests on many individuals. The unit of loudness is the **sone**. By definition, a tone of 1000 Hz of *loudness level* 40 phon has *loudness* of 1 sone. So, a tone of 2 sone sounds twice as loud as a tone of 1 sone. A tone of 6 sone sounds six times as loud compared to a tone of 1 sone, and so on. Table 4.4 lists some values for tones of 1000 Hz. We note that above 40 dB, each 10 dB step in intensity level corresponds to a *factor* of 2 in loudness.

According to the values listed, if we want to double the loudness, from 1 to 2 sone, we must increase the intensity level by 10 dB, meaning an increase in intensity by a factor of 10. If we want to quadruple the loudness from 1 to 4 sone, then we need to increase the intensity from 40 to 60 dB, an increase of 20 dB, which means increasing the intensity by 100 times. The practical aspect of these comparisons is that two voices of about the same pitch, singing together do not sound twice as loud. It may take ten voices to sound twice as loud as one voice. The values listed here apply to 1000 Hz tones. Similar, but quantitatively different, results are obtained for other frequencies as well[6].

In summary:
- **The intensity level (in dB)** tells us about the intensity of the sound, independent of the frequency.
- **The loudness level (in phon)** compares two tones of different frequencies. It tells us how many dB each should have so that the two tones are perceived as equally loud.
- **The loudness (in sone)** tells us how much we should increase the loudness level (phon) of a tone of a *given frequency* if we want to make it sound 2, 3, 4, …

Table 4.4. Loudness for various intensity levels for tones of 1000 Hz.

Intensity level (dB)	Loudness (sone)
40	1
50	2
60	4
70	8
80	16

---

[6] See discussion in section 4.13.3 on how these numbers are calculated.

times louder. Finding the loudness resulting from adding tones of different frequencies is more complicated, and beyond the scope of this book.

4.8 Just noticeable difference

Related to the discussion of the previous section is the question: what is the smallest *change* in intensity level of a single-frequency sound that one can perceive? This is usually referred to as the **just noticeable difference** in sound intensity level. Table 4.5 lists some representative values of the just noticeable difference.

Table 4.5. Values of the just noticeable difference for intensity levels 30, 60, and 80 dB.

Frequency (Hz)	Just noticeable difference in intensity (dB)		
	30 dB	60 dB	80 dB
250	1.2 (32%)	0.5 (12%)	0.4 (10%)
1000	1 (26%)	0.4 (10%)	0.3 (7%)

From the values listed, we note that at low intensity levels (30 dB) the change must be about 1–1.2 dB or 25%–30% (values in second column) to produce a noticeable change in loudness, compared to 0.3–0.4 dB (7%–10%) at 80 dB[7]. The values suggest that our ability to distinguish changes in intensity level near the TOH is not very good. The same actually applies at very high levels (around 100 dB and above). Our ability to distinguish at intermediate sound levels improves, as indicated by the lower values of 0.3–0.5 dB (7%–12%) listed in the last two columns.

4.9 Frequency and pitch

Loosely speaking, the pitch of a sound wave refers to the frequency of that sound. The frequency can be objectively measured in Hertz, and does not depend on the subject. The pitch is the perception of frequency by the individual, and as such, depends on the individual being tested, and how the test is conducted. In distinguishing sound intensity level changes, our hearing cannot perform to better than a few percent. In contrast, our hearing is much more sensitive to frequency differences. Table 4.6 lists representative values of the **just noticeable difference** in frequency, for selected frequencies in the audible range. In terms of frequency difference (middle column) the **just noticeable difference** increases with frequency. Note, however, that compared to the values in table 4.5, the % values in the third column indicate that our hearing is much better at distinguishing frequency changes than distinguishing intensity levels.

---

[7] For relation of dB and % values see section 4.14.3.

Table 4.6. Just noticeable difference at various frequencies.

Frequency (Hz)	Just noticeable difference (Hz)	(%)
100	1	1
500	2	0.4
1000	3	0.3
4000	20	0.5

It is interesting to note that when more than one frequency is present, our hearing may perceive additional frequencies, called combination frequencies, that are not included in the sound signal. For example, if we sound two tones, say 220 and 320 Hz, we may hear a tone of 100 Hz. The frequency of the additional tone is the difference $320 - 220 = 100$ Hz, and it is called the **difference tone**. Another interesting situation occurs when we have a sequence of frequencies say 330, 440, 550, and 660 Hz. Note that the frequencies are multiples of 110 Hz; in other words, the frequencies would be the harmonic overtones of a fundamental frequency[8] of 110 Hz. In this case, the listener could perceive the combination as a pitch of 110 Hz, the so-called **virtual pitch or missing fundamental**. This effect works mostly in the low frequency range, under approximately 1 kHz.

In section 3.5, we compared the sound frequency spectrum of guitar strings (see figure 3.11.) What is noteworthy is that the lowest frequency, which we associate with the pitch of the tone, is not the loudest. Yet, because of the virtual pitch, our hearing associates the pitch of the complex tone with the lowest frequency. Another interesting application of the virtual pitch is the perceived bass sound from small loudspeakers and earphones (see section 7.7). Our ears may pick up a sequence like 220, 330, 440 Hz, and perceive a 110 Hz pitch which is not present in the output of the loudspeaker altogether. So, one gets more bass sound from a small loudspeaker or an earphone, which do not have much output in the bass range.

4.10 Structure and function of the human ear

The basic function of the ear is to convert acoustic pressure to an electrical signal that is transmitted to the brain. In a broad sense this function is similar to what a microphone does, that is, to convert the pressure pattern of acoustic waves to an electrical signal. Of course, the structure and function of the ear are far more sophisticated and complex, and the major difference is that the microphone produces an electric voltage that is a replica of the pressure wave, while the ear produces a series of pulses of approximately the same height. If the pressure doubles, the microphone voltage doubles. In the case of the ear, if the pressure doubles, the ear produces more pulses. Figure 4.3 shows the essential parts of the mechanism that transmits the pressure signal to the inner ear, where it is converted into electrical pulses. First of all, the sound wave is collected by the ear lobe, which guides it into

---

[8] See section 3.2.

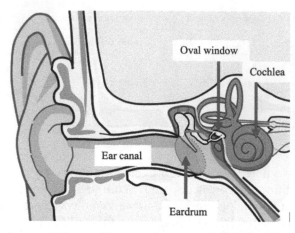

Figure 4.3. Schematic diagram of the ear. The acoustic pressure is transmitted from the eardrum to the cochlea, where it is converted to an electrical signal. Attribution: Anatomy_of_the_Human_Ear.svg: Chittka L, Brockmann derivative work: M Komorniczak -talk-, CC BY 2.5, via Wikimedia Commons.

the ear canal. Keep in mind that though that a good part of the sounds that we hear, especially our own voice, comes from acoustic waves travelling through our bones. These sounds skip this stage, but they are eventually processed in the same way by the inner ear. The ear canal is roughly a cylindrical tube approximately 3 cm long and 1 cm wide. At the inner end of the canal is the eardrum, a 0.1 mm thick membrane. The backside of the eardrum is linked by a system of three bones to another membrane, the so-called oval window. The area of the oval window is about 10 times smaller than the area of the eardrum and the material is thinner and more flexible. The backside of the oval window is linked to the so-called basilar membrane, which is enclosed in the cochlea, a spiral bony structure filled with fluid. The fluid in the cochlea is separated by the basilar membrane, which stretches along the cochlea. The side of the basilar membrane closest to the oval window is narrow and stiff, and becomes wider and softer as we move away from the oval window. The total length of the basilar membrane is about 3 cm. One side of the basilar membrane is covered with the so-called hair cells, roughly 20 000 of them. The hair cells connect to the nerve endings, which combine to form the auditory nerve (yellow feature connected to the right side of the cochlea in figure 4.3) which carries the information to the brain.

When an acoustic wave from the surroundings comes into the ear, it sees a significant change in impedance and, as discussed in section 2.4, this means a strong reflection back to the surroundings. A significant function of the ear lobe is to reduce the reflection by making the change in impedance more gradual. This is what we called *impedance matching* in section 2.4. When the wave reaches the end of the canal, part of the energy reflects back and part of the energy sets the eardrum to vibration. The three-bones that link the ear drum to the oval window act as a lever, that not only transmits the vibrations to the oval window, but in the process, amplifies the pressure on the oval window. The vibration of the oval window sets the fluid in the cochlea into motion, which in turn, sets the basilar membrane into a wavy motion, like a waving flag. As this happens, the hair cells begin to bend, and this bending is more when the flow of the fluid is stronger. This happens when the vibration of the oval window is stronger. As the hair cells bend, they act as valves that can open and allow some electric charge to move towards the nerve endings to which the hair cells are attached. This motion of electrical charge triggers the nerve (makes it 'fire') which sends pulses of electric voltage to the brain. When the oval window vibrates strongly, it causes more bending of the basilar membrane, and eventually more pulses to the brain. This explains how our brain receives information about the intensity of the sound. It is a bit more complicated to explain how the brain gets the information regarding the frequency content of the acoustic wave. One possible way of processing the frequency content could be by timing the passage of highs and lows through the basilar membrane. Another way is suggested by the properties of the basilar membrane. Recall that near the oval window the membrane is narrow and stiff, implying that shorter wavelengths (higher frequencies) may interact more effectively with the membrane near the oval window. As we get further away from the oval window, the membrane gets wider and softer, and that would be the place where the longer wavelengths (lower frequencies) are detected. In this process, our brain understands the frequency by knowing the place (on the basilar membrane) of the hair cells that caused the nerves to fire. In other words, our brain gets information about the frequency content because the different frequencies interact with different places on the basilar membrane. Either way, the output from the ear that is sent to the brain contains information about the amplitude of each frequency contained in the sound signal, which is what we called the amplitude spectrum in section 3.3.

4.11 Critical bands

As described in the previous section, the processing of frequencies occurs in the inner ear. In terms of musical sound, an important question is what happens if we have two pure tones sounded simultaneously. We have already discussed the case of beats (section 1.4) where the frequencies of the *two tones differ by a few Hz*. Beats are clearly noticeable if the beat frequency (which is equal to the difference between the frequencies of the two pure tones) is less than about 10–15 Hz. If the difference in *frequencies is very large, say a few hundred Hz*, then the perception of one frequency does not seem to be affected by the presence of the other frequency. If the difference in frequencies is in between, then perception of one tone is affected by the presence of the other. This interaction is attributed to the existence of the so-called **critical bands**, which essentially define a frequency range around each tone. If the difference between the two frequencies is smaller than the width of the critical band, then the two frequencies interact. The width of the critical band depends on the frequency range in question. For example, for frequencies below 500 Hz, the width of the critical band is about 100 Hz. For frequencies around 1000 Hz, the width of the critical band is about 150 Hz, and for frequencies about 3000 Hz, the width is 300 Hz. The origin of critical bands is beyond the scope of our discussion. For our purposes, we will use the concept of critical bands to describe two important effects, which are **masking** and **roughness**.

4.11.1 Masking

When the frequencies of the two tones are close, then some **masking** occurs. What this means is that if we check the threshold of hearing (TOH) at say 250 Hz (the testing frequency) with the simultaneous presence of a 200 Hz tone (the masking frequency) then the TOH for the 250 Hz tone will be higher than the 10 dB value listed in table 4.2. On the other hand, if we check the TOH for 250 Hz (the testing frequency) with the simultaneous presence of a 1000 Hz tone masking frequency, the TOH we find for 250 Hz tone will be 10 dB, which is the same as it was without the presence of the 1000 Hz masking frequency. For masking to occur, the difference between the frequencies of the two tones must be less than the bandwidth of the critical band around the frequency in question. For example, around 200 or 250 Hz, the bandwidth of the critical band is around 100 Hz. This means the 200 Hz tone can mask the 250 Hz tone, and vice versa, because the frequency difference between them is 50 Hz, less than the critical bandwidth. On the other hand, if the masking frequency is 1000 Hz, then, the difference between tones is 1000−250 = 750 Hz, which is more than the critical bandwidth around 1000 Hz, which is 150 Hz. The same applies if the masking tone is 250 Hz, and the testing frequency is 1000 Hz. In addition, the extent of masking depends on the intensity of the masking frequency. Higher masking intensity increases the TOH for the test sound. What is also interesting is that a test frequency is more effectively masked by a lower frequency. In other words, it is more effective to mask a 250 Hz tone using a 200 Hz masking tone rather than a 300 Hz masking tone.

4.11.2 Roughness

For frequency differences slightly larger than what would lead to perceptible beats, the combination of the two tones produces an unpleasant **roughness**. As the frequency difference increases the roughness diminishes and finally disappears as the frequency difference gets larger. Exactly how large the difference between the two frequencies should be for this roughness to disappear depends of the frequency range in question. The dependence on the frequency difference can be understood in terms of the critical band, which essentially defines a frequency range around each tone. For example, for tones around 200 Hz, the critical band extends from about 100 up to 300 Hz. So, if we play *simultaneously* a tone that is within the critical band of the 200 Hz tone, say a tone of 220 Hz, we will perceive a rough tone of (200+220)/2 = 210 Hz. If a tone of 200 Hz is simultaneously sounded with a tone of 350 Hz (outside the critical band) then we perceive two separate tones. For frequencies around 1000 Hz, the critical band extends from about 850 up to about 1150 Hz, meaning that the combination of 1000 and 1030 Hz is a rough tone of 1015 Hz. Again, if the 1000 Hz tone is sounded simultaneously with a tone of 700 Hz (outside the critical band) then we perceive two separate tones.

Audio 4.1. Sum of a pure tone of 440 Hz, and a tone of variable frequency. Available at https://iopscience.iop. org/book/978-0-7503-3539-3.

Audio 4.1 should help illustrate the above concepts. The duration of the audio is 100 s, and it plays two pure tones, Tone 1 and Tone 2, simultaneously. The two tones have equal amplitude. The frequency of Tone 1 is held fixed at 440 Hz. The frequency of Tone 2 is variable. It starts at 440 Hz, and decreases by 1 Hz every second to a final value of 340 Hz. Note that for the first 10–15 s we hear beats. The beat frequency increases, and the two tones fuse into one. After about 15 s, the beats give way to a rough sounding fused tone. This happens for approximately the next 70 s. As time elapses, the frequencies get further apart, and we hear two distinct sounds. For the frequencies used here, the width of the critical band is about 100 Hz. This means that frequency separation between the two tones gets larger than the critical bandwidth only towards the end of the audio. In this audio one may perceive some low frequency sounds, corresponding to the difference between the two frequencies at that instant. What is interesting here is that the amplitude spectrum at any time consists of two frequencies only: the 440 Hz and the variable frequency. So, the perceived lower frequencies are not in the audio signal. They are created in the listener's ear. These are the *difference tones* mentioned in section 4.9 above.

4.12 Hearing, vision, and the role of the brain

In this section, we briefly discuss some of the common features of our hearing and vision. In both cases, the stimulus is a wave in air. Sound waves in air are longitudinal, and light waves are transverse. Both hearing and vision respond to the frequency of the wave, but the frequencies of sound waves are billions of times lower compared to the frequencies of light waves. The range of audible frequencies is from 20 to 20 000 Hz. Comparing the low and the high end, we see that they differ by a factor of 1000. The range of visible frequencies is from 400 to 800 trillion Hz. Comparing the low and the high end, we see that they differ by a factor of two. So, the range of audible frequencies is relatively much wider.

In terms of intensity, our hearing range is about 120 dB, and as explained above (section 4.4) this means that the value at the higher end is about 1 trillion times larger than the value at the lower end. Our vision responds to a slightly wider range of intensities, closer to about 100 trillion, that is, 140 dB. In both cases, the response to intensity is not direct. Doubling the intensity of sound does not double the perceived loudness. Similarly, doubling the amount of light shining on a surface does not double the perceived brightness. Our hearing can pick up small differences in frequencies (section 4.9), but is not as good at picking up differences in intensities (section 4.8). The reverse is true for vision: we can pick up slight differences in light intensity, but we cannot pick up slight differences in frequency (color). Our vision has a unique ability to combine light of different colors (that is, light of different frequencies) and interpret the combination as a different color. For example, all the colors seen on a computer display are generated using red, green, and blue pixels only. This is the basis of the so-called RGB display. For instance, combining green and red pixels can give us a series of yellow-orange hues. Although the underlying mechanism is very different, combining sound frequencies can also produce analogous effects, such as the difference tone and virtual pitch described in section 4.9.

An important common feature is that both systems have two *receivers*. We have **binocular** vision (two eyes) and **binaural** hearing (two ears) which enhance the perception of our surroundings. Our field of view is about 200° wide. Our hearing has a continuous field of 360°. Binocular vision allows us to determine depth. Binaural hearing allows us to determine the direction and motion of the sound source, even when the source is not visible. This is referred to as **localization** of the sound source.

For low frequency sounds, under 1000 Hz, localization relies on the difference between times of arrival to each ear. If the sound source is to our right, the sound arrives to our right ear first, and then to the left ear. For higher frequencies, localization relies on the difference in intensity. If the source is to our left, then the sound intensity at our left ear is higher. The localization cues given by binaural hearing are often used in filmmaking. Using a stereo sound system, the sound engineer can progressively shift the volume of a sound source, for instance, a person speaking, from the right to the left channel of the stereo system. The localization cue to the viewer is that the speaker is moving, even if the person speaking is not visible. In the same way, shifting the volume of the sound from a car engine from the right to the left channel of a stereo is a cue that the brain of the viewer interprets as motion from right to left.

Our hearing handles less information than our vision. The ear receives a time sequence of pressure, while our eyes receive two-dimensional images. This means that the number of sound-detecting cells in the ear, the so-called *hair cells*, is relatively small (in the order of a few tens of thousands) while our photoreceptors inside the eye exceed 100 million. In either case, the bits of information received from our vision or hearing system would overwhelm the brain. This does not happen, because the information is locally processed first (that is inside the ear or eye) and the brain receives a 'summary' of what is important.

The information from the eyes, and the ears, ends up in the brain in the form of groups or trains of electrical pulses. The height of the pulses is roughly constant, and both sides of the brain receive information from each ear (and each eye). The intensity of the perceived sound affects the number of pulses received by the brain. So, if the intensity of sound (or light) goes up, then the pulses arrive at a faster rate.

This fact can help us understand why our hearing (and vision) cannot follow fast variations in intensity. In section 4.11 it was noted that our hearing cannot perceive beat frequencies higher than about 10–15 Hz. In other words, our hearing cannot follow ups and downs in intensity if they happen faster than 1/15 to 1/10 s or 0.07 to 0.1 s. Suppose that the sound intensity changes back and forth from 20 to 30 dB, and for the sake of argument, let us assume that the 20 dB sound causes a train of 20 pulses and that the 30 dB sound causes a train of 30 pulses. How far apart should these two trains be, to remain distinguishable ups and downs? The perception of beats (which are perceivable up to 10–15 per second) tells us that the ups and downs should be at least 0.07–0.1 s apart. In other words, if the ups and downs of intensity get any closer, we will perceive a steady sound[9]. This is an interesting fact: the transmission of information to the brain works as long as things don't change very fast. Which leads to the conclusion that since the audio frequencies (20–20 000 Hz) are too fast for the brain to handle, the information about frequency must come in a code form to the brain. In other words, the frequency analysis of the sound is done at the inner ear. So, the acoustic nerve carries electrical pulses, and these pulses contain, in a code that the brain can understand, information about the intensity (amplitude) for each frequency contained in the sound signal.

4.13 Further discussion

4.13.1 The equal loudness contours

ISO is the International Organization for Standardization. Graphs prepared by ISO representing the loudness-intensity relation based on many independent tests are shown in figure 4.4.

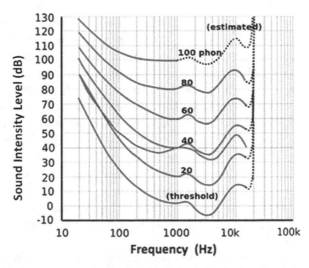

Figure 4.4. Equal loudness contours. Attribution: Lindosland, Public domain, via Wikimedia Commons.

---

[9] Vision behaves roughly in the same way. This is how images fuse in animation and motion pictures.

The subjects were 18–25 years old. The sound source produced pure tones and was directly in front of the listener, so the test was using both ears (binaural listening). In the graph, we use the intensity level on the vertical axis. Note that the vertical grid lines are not equidistant, and the value of the steps in the horizontal axis (frequency) change by factors of 10. The vertical grid lines to the right of 10 mark the values 20, 30, 40 Hz, and so on. The lines to the right of 100 mark the values 200, 300, 400 Hz, and so on. The red curves indicate equal loudness level, and are shown in steps of 20 phon. The red curve marked 40 shows the dB intensity value that would be perceived as having loudness level equal to 40 phon.

Example: Suppose we have a 1000 Hz tone of intensity level 40 dB. What should be the intensity level of a 100 Hz tone in order to sound as loud as the 1000 Hz tone?

Using figure 4.4, we look at the 40 phon curve, and find the point where it crosses the 100 Hz vertical gridline. For that point, read the dB value on the vertical axis. In our case this value is 62 dB. What this means is that the intensity of a 100 Hz tone should be 62 dB, for it to sound as loud as loud as a 40 dB 1000 Hz sound. This is a difference of 22 dB, which is a factor over 100!

The frequencies at which our hearing requires the least intensity level (meaning, our hearing is the most sensitive) correspond to the lowest values of the red curves, and occur around 3000 Hz (3 kHz). The blue line indicates the older standard values at a loudness level of 40 phon.

4.13.2 Some cues and miscues from hearing

Besides the cues discussed in section 4.12, the frequency of a sound can give us some distance cues. As discussed in section 2.7, for complex sounds traveling through the atmosphere, the high frequency components are attenuated more than low frequency components. This means that a complex sound from a distant source will lose most of the high frequencies, and will sound 'dull' compared to sound from a nearby source. Also, the frequency of a tone, as received by an observer, can be different depending on the motion of the sound source. A most familiar example is the frequency of the sound we receive from the siren of a fire truck. When the siren is approaching, the frequency we hear is higher. When the siren is moving away, the frequency we hear is lower. Of course, for the driver the frequency remains the same. This is an example of the so-called **Doppler effect**. In this case it is the *motion* that affects the frequency reaching us, and if we measure it, it would be a shifted frequency. In other words, the shift we hear is real, and not associated with perception. In the case of light waves, the shift in frequency is extremely small, therefore at ordinary speeds our vision cannot detect the Doppler effect due to the motion of a light source.

A most interesting example of the interplay between the brain and our hearing system is **ventriloquism**, literally meaning **talking from the belly**. Of course, a belly cannot produce speech. A ventriloquist has the skill of talking without moving the lips. So, in the standard setting, the artist holds a puppet, and moves the lips of the puppet in synch with the artist's speech. The viewer's brain associates speech with motion of the lips. Artist and puppet are too close for our ears to locate the sound source exactly, and there we have it: a talking puppet! In **lip-synching** the situation is reversed: here the lips of the artist are moving in synch with the vocals from a playback. Another form of synching is to go through the motions of playing an instrument in synch with a playback, like 'air guitar' and the like.

4.13.3 Relating sone, phon and dB

In section 4.7, table 4.4, we listed values of loudness and intensity level for 1000 Hz. Figure 4.5 shows a graph of the loudness (in sone) versus the loudness level (phon). Note that the vertical axis is logarithmic and the steps are multiplicative factors of two. For loudness levels below 40 phon, the relation is not linear, and not shown here. We could translate the loudness level (phon) into intensity level (dB) using figure 4.5. In doing so, one has to keep in mind the frequency dependence. For the special case of 1000 Hz, the situation is very simple, because the numerical value of the loudness level and intensity level are equal by definition. This was the case used in table 4.4 for simplicity.

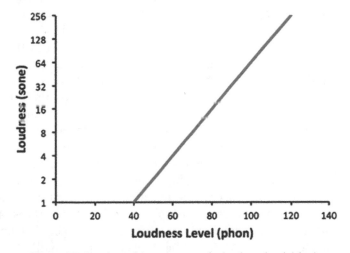

Figure 4.5. Loudness (in sone) versus the loudness level (phon).

4.13.4 The duration of sound waves

The process of recognizing the characteristics of a sound wave begins in the inner part of the ear, and it takes some time to sort out the information content of the sound wave and eventually transmit the signals by the acoustic nerve to the brain. The models that explain the perception of sound are complicated and beyond the scope of this book. Nevertheless, it is plausible that if the duration of the sound wave is short compared to the time needed to complete the perception process, the ear will be unable to completely identify the characteristics of the incoming sound wave. Also, if the duration of the wave is too short, the energy delivered to the inner ear may not be sufficient to properly stimulate the acoustic nerve. The result is that our hearing cannot recognize sound waves of short duration, especially if the intensity is low. Obviously, we need an absolute minimum of at least one cycle (1/frequency) to recognize the frequency of the wave, so the duration of a recognizable frequency cannot possibly be shorter than the time it takes to complete one cycle. For example, if the wave frequency is 1000 Hz, then it takes $1/1000 = 0.001$ s to complete one oscillation cycle. For a frequency of 250 Hz, it takes $1/250 = 0.004$ s to complete one cycle. The number of cycles required depends on the frequency as well. It takes about ten cycles for a frequency of 1000 Hz to be recognized, and in terms of duration this is $10 \times 0.001 = 0.01$ s. This is the shortest duration that a sound of 1000 Hz can have, and be recognizable. For a frequency of 250 Hz, it takes about four cycles. So, the shortest duration a 250 Hz sound can have and be recognizable is $4 \times 0.004 = 0.016$ s.

4.13.5 Sound and music in healing

In ancient times, some people believed that illness was caused by an angry or evil spirit taking over the human body. To cure an illness, this spirit had to be either appeased or expelled from the body. The way this was done usually involved songs, chants, music, and loud noises, alone or in conjunction with offerings, sacrifice and other rituals. Also, music in combination with movement or dance was common in rituals relating to fertility, purification, and the like. Although the belief that illness may have causes other than spirits existed already by the 5th century BC, some of these old rituals survive to our time. The role of music if not as a cure but at least as a means to promote health and wellness was also recognized in ancient times. It is interesting to note that the Greek god Apollo was both the god of music and the god of medicine, probably indicating the belief that both practices are related to some extent.

Music has the power to affect human emotions and state of mind, so it stands to reason that music could be used as a means of reducing stress, lowering heart rate and level of anxiety, all of which benefit our health. Surprisingly, mainstream medicine has only recently recognized the potential of music not just in terms of relaxation, but also as a means of alleviating numerous symptoms and promoting healing. Today, one can hear music in hospitals and medical offices, and playing, listening and dancing to music are commonly prescribed to people recovering from stroke, brain injury and others. Many studies looked for potentially beneficial effects that music may have on people suffering from epilepsy, Alzheimer's, Parkinson's and other diseases. Some very positive, albeit temporary, results have been reported already, and research on this subject is very active.

Another interesting point is the effect of music on our intellectual development. Both Plato and Aristotle believed that music must be an essential part of education. One can hardly argue against the benefits of singing or playing an instrument, which are activities that involve the brain. But how about just listening to music? On that matter, one of the most publicized studies claimed that listening to one particular piece of Mozart's music[10] temporarily improved young peoples' ability to visualize

---

[10] Sonata for two pianos, K 448.

Figure Video 4.1. Singing bowl. Video available at https://iopscience.iop.org/book/978-0-7503-3539-3.

things in three dimensions. The result and its interpretations[11] were verified by some and challenged by other studies. Today it seems well established that music affects the release of chemicals (endorphins, serotonin and other) that reduce pain and produce a feeling of wellness, but what is not clear is *which* element of a given piece of music, and the mechanism of *how* this element makes the process work. To make an analogy, eating an orange is good, because it has vitamin C, which has well established health benefits. So, what is the 'vitamin C' in a piece by Mozart or the Beatles? The challenge here is that there are too many elements in a piece of music.

A seemingly simpler approach relies on some traditional ways that reduce stress and promote wellness using simple sounds produced by percussion instruments, usually metal bowls struck or rubbed by mallets, as shown in video 4.1. This is an ancient Eastern practice that in recent years has become very popular in the Western world, and normally the sounds are part of a meditation or a yoga class. In these 'sound-baths' as they are commonly referred to, the total number of tones sounded during a session is small compared to an orchestral piece of music of equal duration. Reducing the number of variables simplifies the task of relating specific sound characteristics (such as frequency content, duration and volume), to any beneficial outcomes that may materialize.

An even more straightforward approach uses just one frequency, usually in the bass range (see section 4.2). The sound is delivered either directly to a body part, or to the ear. Either way, the sound signal reaches the inner ear, and from there it is transmitted to the brain. One way of delivering the vibration to the body is to apply an electric voltage of the desired frequency to piezoelectric transducers (section 7.4.1) that are attached to the chair used by the patient. Another way is to use tuning forks[12] of the desired frequency. The stems of tuning forks are brought in direct contact with different parts of the body, like the knee, the wrist, and so on. The benefit from direct application may be the result of the mechanical vibration

---

[11] One of them being that listening to Mozart makes us smarter, which is what is commonly referred to as *the Mozart Effect*.

[12] See figure 6.7(b).

transmitted from the stem of the tuning fork to the ailing body part, without going through the hearing system. But it could also result from the processing of the associated sound signal in the inner ear. From there the signal is carried to the brain, where it could initiate synchronized neuron activity, the so-called brainwaves. Recall from section 1.6 that brain wave signals have low frequencies, and that some of these frequencies are associated with particular states of the brain. For instance, frequencies above 30 Hz are associated with a focused and alert state. Frequencies in the range of 4–12 Hz are associated with the relaxed states, and so on[13]. The objective is to use external oscillations to set the desired brain waves to action. This process of setting an oscillator into motion (the brain waves) by making it interact with another oscillator (for example, a sound source) is called **entrainment**. The basic hypothesis here is that if we can initiate brain waves of certain frequency, say in the range of 4–12 Hz, then the brain will *assume* the state associated with the brain waves of that frequency range. In the same way if we can stimulate brain waves above 30 Hz, say 40 Hz, the brain will *assume* the focused and alert state. Although the exact mechanism of how this can work is not clear, positive results have been reported using this procedure on patients with various conditions.

4.14 Equations

4.14.1 Sound pressure level (SLP) scale

We can use the equations from section 2.9.2

$$p_{rms} = \sqrt{\rho c I}$$

to find the root mean square (rms) pressure of the acoustic wave corresponding to the reference sound intensity, which is 10^{-12} W m^{-2}. From table 2.2 we have the density $\rho = 1.2$ kg m^{-3} and the speed of sound $c = 343$ m s^{-1}. Substituting we have:

rms of reference sound pressure $= 20 \times 10^{-6}$ Pa.

From section 4.3 we know that the average intensity at the threshold of pain is 10 W m^{-2}

We can repeat the calculation to find the rms pressure corresponding to the threshold of pain. The result is:

rms of pressure at threshold of pain $= 64$ Pa

In getting the sound intensity level dB, we use the logarithm of the ratio of the average intensity to the reference sound intensity. We know that for any number x, $\log x^2 = 2\log x$. We can use this property of logarithms, and the fact that in terms of the rms pressure the average intensity is given by $I = p^2/2\rho c$ to find:

$$\log(I/I_o) = \log (p/p_o)^2 = 2 \log(p/p_o)$$

---

[13] Our hearing does not pick up frequencies below 20 Hz. Experiments at lower frequencies use rhythmical patterns, such as a sound that changes in volume four times per second. An easy approach is to use the beats (see section 1.4) resulting from the combination of any two audible tones that differ in frequency by 4 Hz.

in other words:

$$10 \log(I/I_o) = 20 \log(p/p_o)$$

where I_o the reference sound intensity (10^{-12} W m^{-2}) and p_o is the rms of reference sound pressure (20×10^{-6} Pa). This last equation tells that we can use a decibel scale involving the rms values of pressure rather than the average intensity. This is the **sound pressure level** or **SLP** for short. The following examples will illustrate the relation in terms of dB.

Example 1: What rms pressure corresponds to 60 dB intensity level?

From the above equation we have $60 = 20 \log (p/p_o)$, or $\log (p/p_o) = 3$, meaning that $p/p_o = 10^3$ or $p = 10^3 \, p_o$ so $p = 10^3 \times 20 \times 10^{-6}$ Pa = 0.02 Pa.

Example 2: What is the SPL of an acoustic wave of rms pressure of 1 Pa?

$$20 \log (1/20 \times 10^{-6}) = 20 \log (50000) = 20 \times 47 = 94 \text{ dB}.$$

Example 3: What is the sound intensity level of the acoustic wave of 1 Pa?

$10 \log(I/I_o) = 20 \log (p/p_o) = 94$ dB (we used the result from the previous example)

So, the numerical value of SPL is the same as the intensity level in dB.

4.14.2 Converting dB to ratios and ratios to dB

The dB unit allows us to compare two intensities, using the logarithm (base-10) of their ratio.

Example 1: If $I_1 = 120$ (W m^{-2}) and $I_2 = 40$ (W m^{-2}) the dB value is

$$10 \times \log(I_1/I_2) = 10 \times \log (120/40) = 10 \times \log (3) = 10 \times 0.477 = 4.77 \text{ dB}.$$

In other words, I_1 is 4.77 dB higher than I_2. Here we used I_2 as reference. We could have used I_1 as reference:

$$10 \times \log(I_2/I_1) = 10 \times \log (40/120) = 10 \times \log (1/3) = 10 \times (-0.477) = -4.77 \text{ dB}.$$

In other words, I_2 is 4.77 dB *lower* (because of the '−' sign) than I_1. If we know the dB value, we can find the ratio, by inverting the logarithm.

Example 2: If I_1 is higher than I_2 by 2 dB, then by the definition of the dB

$$10 \times \log(I_1/I_2) = 2 \text{ dB or}$$
$$\log(I_1/I_2) = 2/10 = 0.2 \text{ so,}$$
$$(I_1/I_2) = 10^{0.2} = 1.62$$

We now know the ratio. If we know the value of the reference (for example, if we know I_2) then

$$I_1 = 1.6 \times I_2.$$

Or if we know I_1, then

$$I_2 = I_1/1.6.$$

The essential point is that the dB is a relative scale. We cannot get from the dB the values of *both* intensities; we can only get their *ratio*. Of course, if we know the ratio, and one of the two intensities (the 'reference' which is usually the denominator in the ratio) then we can find the other (the numerator). Negative dB values indicate that the denominator is larger than the numerator, meaning that the ratio is less than one.

4.14.3 Converting difference in dB to % difference

Here we use % in terms of the reference intensity, that is

$$\% \text{ difference} = 100 \times (I_1 - I_2)/I_2 \text{ or } \%\text{difference} = 100 \times ((I_1/I_2) - 1)$$

So, to convert dB to percentage, we first find the intensity ratio (I_1/I_2); then subtract 1 from the ratio; then multiply times 100.

We will use the values in table 4.5 as examples.

Example 3: A difference of 0.5 dB means that $10 \times \log(I_1/I_2) = 0.5$ or $I_1/I_2 = 10^{0.05} = 1.12$.

So, $(1.12-1) \times 100 = 12\%$.

Example 4: Difference of 1 dB means that $10 \times \log(I_1/I_2) = 1$ or $I_1/I_2 = 10^{0.1} = 1.26$.

So, $(1.26-1) \times 100 = 26\%$.

4.14.4 Other uses of dB units

Use of the dB is not limited to ratios of sound intensities; it is used in the fields of electronics, communications, optics, to mention just a few. In chapter 7 we will use dB to measure power output, for example in audio amplifiers. Recall that dB is essentially a logarithm of a ratio. In the case of the amplifiers, we can use the ratio of the output to the input power. We can also use the ratio of the output to the input rms voltage. In this case, as in the case of sound pressure, we need to use the definition $20 \log (V_{out}/V_{in})$. It is also common to adopt a reference voltage to form the ratio, as we have done with the reference sound intensity. For example, in some instances a voltage is given in dB using 1 V as reference. In this case the units are in dBv. The 'v' is added to remind us that 1 V was used as reference.

4.15 Problems and questions

1. (a) What is the difference between sound power and sound intensity?
    (b) What is the difference between sound intensity and sound intensity level?
    (c) If two pure tones have the same loudness, does it follow that the two tones have the same loudness level?

2. From table 4.1 we see that the sound intensity level of a whisper, a lawnmower and a jet engine at 10 meters are 30, 90 and 120 dB, respectively.
   (a) Does this mean that a whisper plus a lawn mower (30 + 90 = 120) have the same intensity level as a jet engine?
   (b) From the intensity ratios, we see that the intensity of a whisper is 1000 times higher than light breathing, meaning that it takes 1000 people breathing lightly to produce the same intensity level as one person whispering. Using the same logic, find how many lawnmowers it takes to produce the same intensity level as a jet engine.
   (c) What intensity ratio corresponds to intensity level of 40 dB?
   (d) Suppose we have 10 lawnmowers running. What is the intensity level of their combined sound in dB?

3. Suppose that we have two equidistant sound sources producing pure tones of frequency 30 and 1000 Hz.
   (a) If the two sources are adjusted to output an intensity level of 5 dB each, which source will emit more power?
   (b) If the two sources are adjusted to output an intensity level of 5 dB each, which source will sound louder?
   (c) Do sound intensity and loudness level mean the same thing? If not, then explain the difference.
   (d) If the two sources are adjusted so that the perceived loudness level from each source is 40 phons, which source will sound louder?
   (e) How much should the intensity level of the 1000 Hz source be increased in order to make it sound twice as loud?

4. For pure tones of about 1000 Hz, the just noticeable difference in frequency is about 3 Hz.
   (a) Suppose that one hears a tone of 1000 Hz, followed by a tone of 1002 Hz. Would that person be able to tell the difference in pitch?
   (b) Suppose that the two tones are sounded together. Would one be able to tell that the two tones do not have the same pitch?
   (c) What is different between the situations described in parts (a) and (b) of this question?

5. We have two pure tones of sound intensity level 20 dB. One tone has frequency 30 Hz and the other tone has frequency 250 Hz.
   (a) Which of the two tones is below the threshold of hearing?
   (b) Which of the two tones sounds louder?
   (c) Which of the two tones has the higher sound pressure level?

6. (a) When you click your teeth, keeping your mouth closed, you hear a clearly audible sound. But if you try to record the sound you make when clicking your teeth, you will probably capture very little or no sound at all. Can you explain why?
   (b) When you hear your voice form a recording, it sounds very different to you. Can you explain why?

IOP Publishing

The Physics of Sound Waves (Second Edition)
Music, instruments, and sound equipment
Panos Photinos

Chapter 5

Musical scales and temperament

5.1 Introduction

The range of audible frequencies is essentially a continuum, a lot like the colors of the rainbow. So, in principle one could use any combination of frequencies simultaneously or successively, in a similar way that an artist uses colors. However, the results of these approaches may not be pleasing to the ear or the eye. Traditionally, music is based on a set of tones of interrelated frequencies and with rules on how these tones should be combined together to produce a pleasing result. For example, playing the same note for too long, or playing together adjacent notes on a keyboard, is generally unpleasant. It is clear however that different cultures developed their own music, based on different choices of tone relations, instruments, rhythms and so on. In other words, different cultures developed different aesthetic criteria in music, which are not difficult to recognize. For example, most people can easily recognize a piece of Latin American or Indian music. A closer look shows that even within a single culture, the music is not something monolithic that follows a strict set of rules or norms, but rather a fusion

doi:10.1088/978-0-7503-3539-3ch5

of traditions that reflect major historical events, and changing socioeconomic conditions. For example, *flamenco* is Spanish music with significant Islamic, Jewish and Romani influence, and *jazz* is a product of the fusion of European and West African musical traditions. Of course, major turning points in music can also happen because artistic creativity, by its nature, tends to bend or break the existing norms and rules, to make way for something new. This is what gave us Tchaikovsky, the Beatles, and Bob Marley, to name just a few.

In music, the term **scale**[1] is used to indicate a set of ascending or descending steps in *frequency*. There are several kinds of scales, some are used widely around the world, and some are more characteristic of particular cultures. In this chapter we will first describe the frequency relations between the steps of the musical scales used in the western world, and then discuss some examples from different cultures. Here we use a mathematician's rather than a musician's approach to music. It seems that music theory is where art and science at times become indistinguishable.

5.2 Keyboard notes

To make the description more specific we will refer to the keyboard. A standard piano has 88 keys. Figure 5.1 shows a section of a piano keyboard. The keys are labeled with their English names[2].

At once we note that the pattern is repeating, and that the black keys are labeled with reference to the white keys. For example, the black key between C and D is labeled as C ♯ (pronounced C sharp) or D ♭ (D flat). We also note that there are 11 different keys (counting black and white) between two successive As. If we think of the keys as steps in a ladder, then there are 12 steps between successive As on the ladder. The same is true for successive Bs, Cs, and so forth. It is common to refer to the C near the middle of the piano keyboard (fourth C from left end of the standard piano keyboard, see figure 11.10) as **middle C** or C4. All the keys can be numbered similarly.

Figure 5.2 shows a section of the keyboard starting from C2. Note that the numbering is based on the Cs not the As. This means that the white keys between C4 and C5 are D4, E4, F4, G4, A4, and B4.

The black keys are numbered following the same scheme. To describe the direction of movement on the keyboard we will use the term **ascending** to indicate the direction from left to right (which is the direction of increasing tone pitch) and **descending** to indicate the direction from right to left. Note that some of the white keys are separated by a black key (like C4 and D4 in figure 5.2) while others are not separated by a black key (like B3 and C4). We say that C4 and D4 are one **whole tone** apart (or two steps) and B3 and C4 are half tone, or one **semitone** apart, that is, one step apart[3].

---

[1] From Latin *scala*, meaning staircase.

[2] Many countries use the **solfège** system, corresponding to C(do), D(re), E(mi), F(fa), G(sol), A(la), B(si).

[3] It is interesting to note that about three centuries ago, in the times of Bach and Mozart, white keys were black and black keys were white.

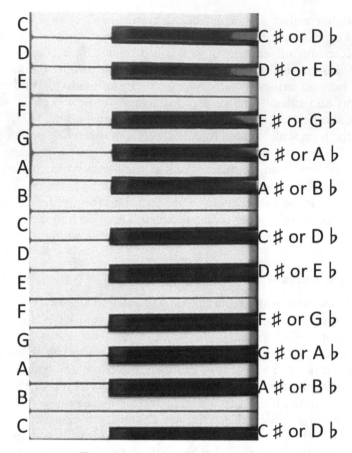

Figure 5.1. A section of a piano keyboard.

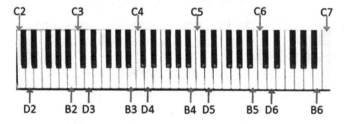

Figure 5.2. A section of the keyboard starting from C2.

5.3 Major and minor scales

If we ascend (or descend) by 12 successive semitones starting from any key, we are playing the **chromatic scale**. The chromatic scale is not commonly used in popular or classical music. A more common scale starts with a C (like C4) and ends with the nearest C (like C5 or C3), and includes in sequence all the white keys. This scale is called the **C-major scale**. In terms of whole tones (W) and semitones (S), the C-major scale follows the pattern:

C-major scale

Note:	C		D		E		F		G		A		B		C
Separation:		W		W		S		W		W		W		S	

In terms of semitone steps, we have:

2, 2, 1, 2, 2, 2, 1

In the C-major scale, we have *seven* distinct notes (C-D-E-F-G-A-B) and the scale is called **heptatonic**[4]. The C-major scale is also a **diatonic** scale, meaning that between the two successive Cs, the pattern includes two semitones that are separated by at least one whole tone. There are many possibilities for heptatonic diatonic scales. For example, starting with an A (like A4) and ascending (or descending) to the nearest A (like A5) and playing in sequence all the white keys, we have the **A-natural minor** scale. In terms of whole tones (W) and semitones (S), the A-natural minor scale follows the pattern:

A-natural minor scale

Note:	A		B		C		D		E		F		G		A
Separation:		W		S		W		W		S		W		W	

In terms of semitone steps, we have:

2, 1, 2, 2, 1, 2, 2

Comparing the two scales above, we see that they both have the *same keys*, but the position of the two semitones relative to the starting note are different, and as a result, the two scales sound very different. The A-natural minor scale is also a heptatonic (has seven distinct notes) diatonic (has two semitones separated by at least two whole steps) scale.

Generally, minor scales[5] may sound 'sad'; for example, the *funeral march* in the third movement of Mahler's Symphony No. 1. Major scales sound 'happy', like the nursery rhyme Brother John (Frère Jacques). It is interesting to note that Mahler's funeral march is a conversion of the nursery rhyme to a minor scale!

In an oversimplified picture, we can view a musical tune as a timed sequence of steps of various sizes. For example, *Twinkle Twinkle Little Star* in the C-major scale would look like:

Note:	C	C	G	G	A	A	G
Lyric:	twin-	kle	twin-	kle	lit-	tle	star
Separation in semitones:		ascend 7		ascend 2		descend 2	

---

[4] *Hepta* means seven in Greek.
[5] See brief discussion of harmonic and melodic minor scales in section 5.8.2.

We can get a different tune, also in the C-major scale, using a different sequence of steps:

Note:	C	C	C	G	A	A	G
Lyric:	Old	Mac	Don-	ald	had	a	farm
Separation in semitones:			descend 5	ascend 2		descend 2	

Depending on the range of tones that a singer's voice can produce, it may be necessary to start a song from a different note. For example, *Twinkle Twinkle Little Star* can be played as follows:

Note:	F	F	C	C	D	D	C
Lyric:	twin-	kle	twin-	kle	lit-	tle	star
Separation in semitones:		ascend 7		ascend 2		descend 2	

Here we **transposed** the song from the C-major scale to the F-major scale. In terms of whole tones (W) and semitones (S), the F-major scale follows the pattern:

F-major scale

Note:	F		G		A		B♭		C		C		E		F
Separation:		W		W		S		W		W		W		S	

Note that the separation is the same as for the C-major scale, which means that in terms of semitone steps we still have:

2, 2, 1, 2, 2, 2, 1.

As will be discussed in the next section, in the modern system of tuning, the semitone steps are all the same. The equivalence of steps allows us to construct major scales (and minors scales) starting from any note. In other words, we can have 12 major scales, and 12 natural minor scales.

Pattern of semitone steps for all the major scales:	2, 2, 1, 2, 2, 2, 1
Pattern of semitone steps for all the natural minor scales:	2, 1, 2, 2, 1, 2, 2

Each scale is named after the first note in the scale, which is referred to as the **tonic** of the scale. Figure 5.3 shows three major and three minor scales. The tonics here are

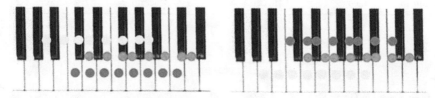

Figure 5.3. Three major and three natural minor scales: A (yellow dots), C (red dots), and D (green dots).

A (yellow dots), C (red dots), and D (green dots). Using the patterns of semitone steps, the reader could construct the major and minor scales for any tonic.

5.4 Frequency relations and intervals

In chapter 3 we found that even a single note from a musical instrument contains a number of frequencies; the fundamental and the overtones. Here the term *frequency* will refer to the frequency of the fundamental of the tone in question. A list of the fundamental frequencies of the 88 pianos notes is given in appendix C. Table 5.1 lists a set of four tone frequencies as used in the standard piano tuning. The fifth column of the table shows that the difference between the frequencies of F4 and E4 is not the same as the difference between F5 and E5. This tells us that the difference between frequencies is not a consistent measure for specifying the frequency spacing of the tones. Looking at the sixth column, we see that the frequency ratio of F4 and E4 is the same as the frequency ratio of F5 and E5. This means that the **frequency ratio** between tones is suitable measure to specify the frequency spacing of the tones.

Table 5.1. Frequency difference and frequency ratio for two tones separated by one semitone.

Note	Frequency (Hz)	Note	Frequency (Hz)	Difference F−E (Hz)	Ratio F/E
E4	329.63	F4	349.23	19.60	1.059 46
E5	659.26	F5	698.46	39.20	1.059 46

The **interval** between two tones is the ratio of the frequencies of the two tones. For the example in table 5.1, the interval between F4 and E4 is 1.059 46, and is exactly equal to the interval between F5 and E5. In forming the ratio, we use the higher frequency in the numerator. In fact, it is 1.059 46 for any consecutive Es and Fs on the keyboard. Therefore, if we know the frequency of E2, then multiplying that frequency times 1.059 46 gives us the frequency of F2. More generally, if we know the frequency of any note (black or white key), then multiplying that frequency times 1.059 46 gives the frequency of the next note (in the ascending direction) on the keyboard, which may be a black or a white key. This relation between tone frequencies is the basis of the modern tuning system, which is discussed next.

5.5 The equal temperament scale

An ordinary piano is tuned so that the interval between any two adjacent notes (including black keys) is the same. This tuning is called the **equal temperament tuning** or **equal temperament scale**. Note that E and F are adjacent (there are no white or black keys between them) and according to table 5.1 the interval between them is **1.059 46**. In the equal temperament tuning, the interval of all semitones must be the same. It follows then that all semitone intervals in the equal temperament scale are equal to 1.059 46. Using the values in table 5.1, a simple calculation shows that the frequency ratio (that is, the interval) between E5 and E4 is 2, and that the frequency ratio between F5 and F4 is 2 also. The same is true for all consecutive Cs or Ds or

As, and so on. So, if we know the frequency of E3, multiplying that frequency times 2 gives us the frequency of E4. Frequency ratio = 2 means that the frequency of E5 is twice the frequency of E4, and defines a very important interval, which is the **octave**. Since we know that the interval between semitones is 1.059 46, all that we need to build a scale in terms of frequency is to choose the frequency of one key. The common choice is to assign to **A4 the frequency of 440 Hz**. So, if the frequency of A4 is 440 Hz, the frequency of A4♯ = 440 × 1.059 46 = 466.16 Hz. The next note is B4, with 466.16 × 1.059 46 = 493.88 Hz, and so on.

Table 5.2. Interval and frequency relations for the C4-major scale[6].

Musical note	Freq. (Hz)	Semitone steps from previous note	Interval from C4	Name of interval
C4	261.63		1.000	Unison
D4	293.66	2	1.122	Major second
E4	329.63	2	1.260	Major third
F4	349.23	1	1.335	Fourth
G4	392.00	2	1.498	Fifth
A4	440	2	1.682	Major sixth
B4	493.88	2	1.888	Major seventh
C5	523.25	1	2.000	Octave

Table 5.2 lists the frequencies of the white keys in one octave in the equal temperament tuning, ascending from the middle C. As noted earlier, this is the C major scale. The fourth column lists the intervals from C4, which is the frequency of the note in the second column, divided by the frequency of C4. The intervals have proper names as well, which are listed in the fifth column. For example, if we ascend from C4 by two semitones, we have the major second of the C-major scale, which is D4. If we ascend another two semitones (that is a total of four semitones from C4) we have the major third of the C-major scale, which is E4, and so on. The proper names in the fifth column arrange the notes in a hierarchy with respect to the tonic of the scale, which in the case of table 5.2 is C4. Similar intervals are defined for all major and minor scales.

As expected, the pattern of semitone steps in the third column is identical to the semitone pattern listed at the end of section 5.3. One could make a similar list for the C4 natural minor scale. The difference will that the minor third, sixth and seventh would be one semitone flat, compared to the corresponding major interval. The second[7], fourth and fifth are the same in the major and natural minor scale.

---

[6] The fourth and fifth are sometimes referred to as perfect fourth and perfect fifth, respectively. This will be discussed in section 5.6.

[7] The term 'minor second' is used to indicate one half tone interval from the tonic, that is C4♯ in the example of table 5.2. The minor second is *not* the second note in the minor scale. For a given tonic, the second note in the minor scale is the *same* as the second note in the major scale.

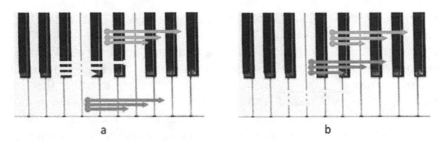

Figure 5.4. (a) Major thirds, the fourths, and fifths for A (yellow arrows), C (red arrows), and D (green arrows) tonics. (b) Natural minor thirds, the fourths, and fifths for the same tonics.

Figure 5.4(a) shows the major thirds, fourths, and fifths for three tonics; which are A (yellow arrows), C (red arrows), and D (green arrows). Figure 5.4(b) shows the minor thirds, the fourths, and fifths for three tonics; which are A (yellow arrows), C (red arrows), and D (green arrows). Note that for a given tonic, the fourths and fifths are the same for the major and the natural minor scales.

If the two tones involved in an interval are played simultaneously, we speak of a **vertical** or **harmonic interval**. If played in sequence, we have a **horizontal** or **melodic interval**. In the same way, three or more notes form a **chord**, which can be played vertically or horizontally. In the discussion that follows we refer to vertical intervals.

5.6 Consonance and dissonance

From ancient times, it was known that some combinations of tones sounded good together, and others did not. While we can only guess how the music of different cultures sounded, we can get a pretty good idea of the intervals used, for example, by analyzing the size and distances between holes in ancient flutes (see sections 6.6 and 6.7). Around 500 BC, Pythagoras discovered the mathematical relation between some intervals that were commonly used at his time. These intervals were probably known by many cultures and for many centuries before Pythagoras. It is said[8] that by some experimental measurements, Pythagoras established that a frequency ratio of 3:2 sounds very pleasant. This interval is known as a **perfect fifth**. He also found that the frequency ratio of 4:3 is also pleasant and this interval is known as a **perfect fourth**. Note that in both cases, as well as other intervals, for example, the octave (ratio 2:1), the frequency relation is a ratio of two **small integer numbers**. Pythagoras did not tell us anything about the frequency itself, obviously, because the measurement of frequency had to wait for over 2000 years. What Pythagoras accomplished was to measure experimentally the frequency *ratios* (that is, the *intervals*) that sounded most pleasant. So, Pythagoras' conclusions apply for any tonic.

Over the centuries, several systems of tuning, which are ways of assigning frequencies to each tone in an octave, were developed starting with ratios of small numbers. Two such scales are discussed in sections 5.9.1 and 5.9.2. What one should

---

[8] We do not have any information directly from Pythagoras. It seems that he did not write anything, and on purpose!

note is that the intervals $3:2 = 1.5$ and $4:3 = 1.333$ are close to the fifth and the fourth in table 5.2, but *not* quite the same.

In Pythagoras' times, the reason why some combinations were pleasant sounding, or **consonant**, when played together simultaneously, was attributed to more or less mystical properties of numbers. In the past two centuries, scientific theories were advanced to explain **consonance**, and its opposite, **dissonance**. In section 4.11.2 we discussed the concept of roughness, and its relation to the critical bands. In this section we will use examples to revisit this discussion in the context of intervals and musical scales[9].

Example 1: The fundament frequency of A4 is 440 Hz. The frequency of A4♯ is 466. The frequency difference is 26 Hz, which is too fast for our ear to pick up as a beat. On the other hand, the width of the critical band in this range of frequencies is about 100 Hz. This means that the two tones are separated by about one quarter of the width of the critical band, in other words, they are too close, and the combination is rough, or dissonant.

Example 2: The fundamental frequency of E4 is 330 Hz. Compared to A4, we have a difference of $440 - 330 = 110$ Hz, which is more than the width of the critical band in this range of frequencies (100 Hz). This means that the combination of E4 and A4 is consonant. In fact, counting steps in the middle column of table 5.2 we see that A4 is 5 semitones higher than E4, which makes A4 the fourth of E4.

Example 3: The fundamental of F4 is 349 Hz. Compared to A4 we have a frequency difference of 90 Hz, which is just below the width of the critical band (100 Hz). So, the roughness is imperceptible. Again, counting steps in the middle column of table 5.2 we see that A4 is 4 semitones higher than F4, which makes it the major third of F4.

As the next example shows, the situation is very different with the lower notes.

Example 4: The fundamental frequency of E1 is 41 Hz, and that of A1 is 55 Hz. A1 is still the fourth of E1, but the frequency difference, 14 Hz in this case, is much smaller than the width of the critical band (about 90 Hz in the low frequency end) so the combination is rough. In the same way, the fundamental frequency of F1 is 44 Hz, and that of A1 is 55 Hz. The frequency difference is 11 Hz, which is much smaller than the width of the critical band, which means that the combination is rough. We note that the frequency difference is low enough to result in audible beats. In the same way one can show that D1 and A1 (a fifth) is also dissonant. The problem here is that the notes at the low frequency end are too close, and playing them simultaneously will result in roughness, and audible beats, in short, we have dissonance.

---

[9] The frequencies were rounded to the nearest integer for simplicity.

5.7 From dissonance to consonance

From the intervals in the equal temperament tuning listed in table 5.2, we see that the fifth and the fourth (1.498 and 1.335, respectively) are very close to the perfect fifth (3:2 = 1.5) and perfect fourth (4:3 = 1.333). The equal temperament interval for the major third (1.260) shows a larger deviation from the simple ratio 5:4 = 1.250, so, the major third in the equal temperament tuning was considered dissonant a couple of centuries ago. Some Baroque aficionados feel that the fourth and the fifth of the equal temperament scale are dissonant as well, and some feel that the equal temperament scale altogether is dissonant. Be that as it may, it is reasonable to assume that music would more likely sound harmonious if it includes consonant intervals, which evoke relaxation and calm, while dissonant intervals evoke tension. This was the common understanding in the polyphonic music of the renaissance, with many voices singing different tones simultaneously, and even minor dissonance was unacceptable. As it happened over the centuries, intervals that were considered dissonant found their way into major works. A dissonant vertical interval is not necessarily unusable, and composers use the succession of consonance and dissonance to produce emotionally rich and powerful music. As mentioned in section 5.1, it is in the nature of artistic creativity to bend or break rules. The switch of dissonance from *unacceptable* to *enriching* may be one example of how artistic creativity works.

5.8 Other scales

5.8.1 Modes

Starting from C we can ascend the white keys to the next C, and get the familiar C-major scale. We can also start from D and ascend to the next D. The result is another heptatonic, diatonic scale. We could do the same starting from any of the white keys, and get another heptatonic diatonic scale. Proceeding in this manner, we can have a total of seven heptatonic, diatonic scales. These scales are called the **modes** of C-major, which are known by names deriving from ancient Greek, as illustrated in figure 5.5. Note that the Aeolian mode is the ordinary natural A-minor scale.

We could actually look at any major scale, for example use the A-major scale shown in yellow dots in figure 5.3, and generate the seven modes of the A-major

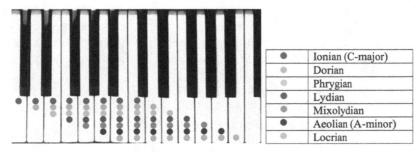

●	Ionian (C-major)
●	Dorian
●	Phrygian
●	Lydian
●	Mixolydian
●	Aeolian (A-minor)
●	Locrian

Figure 5.5. The seven modes of C-major.

scale, or start from D, and use the keys marked by green dots in figure 5.3. In this way we would generate the seven modes of the D-major scale. So, we have twelve different keys, that is twelve distinct major scales, and we can generate seven modes from each, for a grand total of $12 \times 7 = 84$ distinct heptatonic, diatonic scales. Although some experts distinguish the scale from the mode, the fact is that many musical pieces in jazz, folk, and classical are based entirely on a mode, rather than the ordinary major or minor scales.

5.8.2 Other minor scales

In section 5.3, we discussed the natural minor scale. Other commonly used minor scales are the **harmonic minor** and the **melodic minor**.

The pattern of semitone steps for all the natural minor scales:	2, 1, 2, 2, 1, 2, 2
The pattern of semitone steps for all the harmonic minor scales:	2, 1, 2, 2, 1, 3, 1

The patterns above are the same for ascending and descending. Note that the harmonic minor has three semitone steps, and one step of three semitones. This means that the harmonic minor does not follow the rule for diatonic scales, as defined in section 5.3. The pattern for the melodic minor has two semitones separated by at least one whole tone, but the pattern is different on ascending and descending.

Ascending pattern of semitone steps for all the melodic minor scales:	2, 1, 2, 2, 2, 2, 1
Descending pattern of semitone steps for all the melodic minor scales:	2, 1, 2, 2, 1, 2, 2

The descending pattern listed above must be read from right to left. It is interesting to note that the first four intervals for all the minor scales listed above are the same. Having different ascending and descending patterns is not unique to the melodic minor. Many scales used in Arabic, Persian, and other Eastern and North African music have different ascending and descending patterns.

5.8.3 Pentatonic scales

Our previous discussion was based on the heptatonic scale (that is ABCDEFGA), which is the most common. A related scale is the **pentatonic** scale, which uses five distinct tones[10]. Pentatonic scales of various kinds are found in jazz, country music, Native American music, Andean music, Japanese music, and other. One can very simply play a major pentatonic scale, by ascending on the black keys of the piano, starting from F♯.

A **C-major pentatonic scale** would be CDEGAC. In other words, the pentatonic C-major scale derives from the heptatonic C-major by omitting the fourth (F) and seventh (B) tones. In terms of semitone intervals, we have:

---

[10] *Pente* means five in Greek.

Pattern of semitone steps for pentatonic major scales:	2, 2, 3, 2, 3
Pattern of semitone steps for pentatonic minor scales:	3, 2, 2, 3, 2

We note that for the C-major pentatonic scale, the interval from E to G is three semitones, as is the interval from A to C, and there are no single semitone steps.

A **pentatonic scale in A-minor** can be derived from the corresponding heptatonic scale by omitting the second (B) and sixth (F) tone, which means, use ACDEGA. Again, in this particular pentatonic minor there are two three-semitone intervals (from A to C and from E to G) and no single semitone steps. The pentatonic scales do not have single semitone intervals, so they are not diatonic scales.

5.9 Older temperaments

Our discussion in this chapter is based on the equal temperament scale. In this section, we describe two scales that played a significant role in the development of Western music, which are the **Pythagorean** and the **just scales**. We will use the C-major scale, and specify each note by its frequency ratio to the tonic (C in this case). In the following discussion it is important to keep in mind that *ascending* by one interval, say 1.5, from any given note, means multiply the frequency ratio of that note times 1.5. *Descending* by an interval of say 1.333 means divide the frequency ratio of that note by 1.333. The same applies to frequencies. So, *ascending* by one interval, say 1.5, from any given note, means multiply the frequency of that note times 1.5. *Descending* by an interval of say 1.333 means divide the frequency of that note by 1.333.

5.9.1 Pythagorean scale

One of the earlier scales, named after Pythagoras, is constructed using perfect fifths (frequency ratio 3:2) and octaves (frequency ratio 2:1). Table 5.3 lists the intervals of the Pythagorean scale[11]. The second column lists the ratio (for example 3:2) and the third column lists the value (3:2 = 1.5). The fourth column lists the interval between successive notes, so (9:8) /(1:1) = 1.125; (81:64)/(9:8) = 1.125 and so on. The fifth column tells us whether the step is a whole tone (W) or a semi-tone (S), in other words, if there is a black key in between the two white keys or not. The sixth column lists the corresponding interval in the equal temperament scale, and the last column is the difference between the equal temperament and Pythagorean intervals. From the table we see that the ratio of notes separated by a whole step is 1.125, and that the ratio of notes separated by a semi-tone is 1.053. In other words, in the Pythagorean scale, two semi-tones are not equal to a whole tone. That is, 1.053 × 1.053 = 1.109 is different from 1.125. Actually, there is no reason why two semi-tones should equal a whole tone, and as long as we stay on the white keys there is no problem at all. The problem occurs when we decide to use the black keys.

---

[11] The intervals of the Pythagorean scale are derived in appendix D.

Table 5.3. Intervals in the Pythagorean scale.

Note	Interval	Interval	Step	Step	Eq. Temp.	Difference
C	1:1	1.000 00			1.000 00	0
			1.125	W		
D	9:8	1.125 00			1.122 46	0.002 54
			1.125	W		
E	81:64	1.265 63			1.259 92	0.005 71
			1.053	S		
F	4:3	1.333 33			1.334 84	−0.001 51
			1.125	W		
G	3:2	1.500 00			1.498 31	0.001 69
			1.125	W		
A	27:16	1.687 50			1.681 79	0.007 51
			1.125	W		
B	243:128	1.898 44			1.887 75	0.010 69
			1.053	S		
C	2:1	2.000 00			2.000 00	0

For example, suppose that we use the Pythagorean tuning, and assign 440 Hz to A4. According to table 5.3, the frequency of B4 (see figure 5.2) should be a whole tone higher, that is $1.125 \times 440 = 495$ Hz. We can look at the black key between A4 and B4 as an A4 sharp or a B4 flat. In other words, if we ascend one semi-tone from A4 or descend a semi-tone from B4, and get to the same note:

- ascending one semi-tone from A4 means $1.053 \times 440 = 463$ Hz
- descending one semi-tone from B4 means $495/1.053 = 470$ Hz

Which frequency should we use for A4♯ ? So, here is the problem with the Pythagorean scale: we run into inconsistencies if we try to tune the black keys. Of course, there is no reason to stick to the C-major scale. We could use the A-major scale which uses three black keys, and tune the intervals following the values listed in table 5.3. The problem would then arise if we try to play a piece of music in the C-major instead of the A-major scale. This would require retuning of the instruments!

5.9.2 Just scale

The **just scale** is constructed using the both the perfect fifth and the major third. A **triad** comprises three tones. A **major triad** combines a major third and a fifth. For example, starting from C, a major third from C is E, and a major fifth from C is G. The triad C-E-G is called the **C-major triad**. Table 5.4 lists the intervals for the C-major scale in the just tuning. Here we see that the major third, and the perfect fourth, and fifth have the 'right' ratios, that is ratios of small numbers, which are 5:4, 4:3, and 3:2, respectively. This means that the scale includes very consonant intervals. The problem with the just scale is in the whole tone intervals. The notes C and D are

Table 5.4. Intervals for the just scale.

Note	Interval	Interval	Step	Step	Eq. Temp.	Difference
C	1:1	1.000 00			1.000 00	0
			1.125	W		
D	9:8	1.125 00			1.122 46	0.002 54
			1.111	W		
E	5:4	1.250 00			1.259 92	−0.009 92
			1.067	S		
F	4:3	1.333 33			1.334 84	−0.001 51
			1.125	W		
G	3:2	1.500 00			1.498 31	0.001 69
			1.111	W		
A	5:3	1.666 67			1.681 79	−0.015 12
			1.125	W		
B	15:8	1.875 00			1.887 75	−0.012 75
			1.067	S		
C	2:1	2.000 00			2.000 00	0

separated by a whole tone, and the step is 1.125. The notes D and E are also separated by a whole tone, and the step is 1.111. In other words, there are two different whole tone intervals in the just tuning. Also, the semitone interval is 1.067, which means that a step of two-semi-tones is $1.067 \times 1.067 = 1.138$ which does not match either one of the whole tone steps. Again, there is no reason to have the same one whole step interval across the scale, and there is no reason for two semi-tone steps to equal one whole step. But, as in the case of the Pythagorean scale, transposing a piece of music from one scale to another, would require retuning the instruments.

5.10 Noise

In everyday conversation we use the word *noise* to indicate an unpleasant or loud sound. In science noise can indicate unwanted signals in sound equipment (see section 7.6) or a mixture of all frequencies. **White noise** is a random mix of frequencies of equal intensity. Blowing the sound 'fff' would be a close approximation to white noise. **Pink noise** is a random mix of frequencies with equal intensity in each octave. Noise is not necessarily a useless thing. For example, in flutes we essentially start by blowing a more or less random mix of frequencies into a pipe. As described in section 11.4, wind instruments essentially channel the energy of the noise into musical notes.

5.11 Further discussion

5.11.1 Structure of the equal temperament scale

In the modern tuning system (section 5.5) the ratio of a semitone is the same across the keyboard. This means that we have 12 equal semitone intervals in one octave. We also know that the frequency doubles within one octave. It follows then that the semitone interval multiplied by itself 12 times should give 2. In other words, the semitone interval must be the twelfth root of 2. So, in the equal temperament tuning:

$$\text{semitone interval} = \sqrt[12]{2} = 1.05946$$

which is the value used in sections 5.4 and 5.5.

5.11.2 The cent

Using the above equation for the semitone interval we can define an alternative way of measuring intervals between tones, as follows. Taking the logarithm of both sides we find:

$$\log(\text{semitone interval}) = (1/12) \times \log 2$$

We can choose the logarithm of one **semitone interval to equal 100 cents**. It follows then that the interval of two semitones is 200 cents, the interval of three semitones is 300 cents, and the octave (twelve semitones) is 1200 cents.

To convert ratios to cents we use the formula:

$$\text{interval in cents} = 3986 \times \log(\text{ratio})$$

Example: The interval 1.029 30 is $3986 \times \log(1.029\ 30) = 50$ cents.

Note that the interval 1.029 30 is half a semitone, what is called a **quartertone**, because two quartertone steps equal one semitone step, or 1.029 30 × 1.029 30 = 1.059 46.

5.11.3 Quarter tone and microtone scales

A piano can produce only a certain set of frequencies in steps of one semitone. For example, in the standard tuning (see table 5.2) the E4 will produce 329.63 Hz and the F4 will produce 349.23 Hz. As there is no key in between E and F, we cannot produce an intermediate frequency, say 339 Hz. This is not the case for all instruments. A violin, for example (section 11.5.2), can produce all the frequencies intermediate between E4 and F4. As such, the violin can play the perfect fourths and fifths discussed above. In addition, the violin can play steps of half semitones, the so-called half-flat or **quartertones**, which are indicated by the symbol ♭. Such scales have 24, rather than 12 steps within an octave. The larger number of steps allows a larger number of combinations for constructing the equivalent of major and minor scales from each tonic.

In the example of section 5.11.2, we calculated the interval of a quarter tone (ratio = 1.092 30) to be 50 cents. So, descending one quartertone from F4 (349.23 Hz) we get the half-flat F4, or 339.29 Hz, or 50 cents below F4. In practice, the quarter tone interval is not always 50 cents. In Byzantine music, for instance, which is still used in the Greek Orthodox Church, the half-flats usually come with qualifications, such as, *flat* or *not too flat*. The same in Arabic music, Persian music, North African, and others, who use the quartertone scales, the interval is actually *adjustable* at the aesthetic discretion of the performer. In other words, it could be 50 cents, or 40 cents, and so on, depending on the artist's preference.

The traditional Arabic and Turkish systems have about 100 such scales known as **maqamat**[12] which, in the vast majority use quartertones. Audio 5.1 is a recording of a popular maqam, the Rast, in C. The ordinary C major would be CDEFGABC. In the Rast scale, the E and the B are half-flat. Audio 5.2 is a recording of the C-major scale for comparison.

Audio 5.1. Maqam Rast scale in C. Available at https://iopscience.iop.org/book/978-0-7503-3539-3.

Audio 5.2. C-major scale. Available at https://iopscience.iop.org/book/978-0-7503-3539-3.

Microtones and **microtonal scales**, are terms used, to indicate intervals other than the quartertones, like tones flattened by 10, 20, 30 cents and so on, and are popular in modern music.

5.11.4 Second order beats

Figure 5.6(a) shows the waveform resulting from adding two pure tones of equal amplitude, of frequencies 220 and 330 Hz. The corresponding periods are $1/220 = 4.55$ and $1/330 = 3.03$ ms, respectively. Except for the difference in the time scale, this figure is similar to the bottom waveform of figure 1.7, which shows the addition of two pure tones with frequency ratio $3:2 = 1.5$. The pattern repeats approximately every 0.009 s, which corresponds to frequency of 111 Hz. Using the procedures of combining waves (section 1.3) we find that the pattern should repeat

---

[12] Plural of **maqam**, meaning *place* in Arabic.

Figure 5.6. (a) Combining two pure tones of equal amplitude and frequencies 220 and 330 Hz, which corresponds to the perfect fifth. (b) Combining two tones of frequencies 220 and 335 Hz, which is 5 Hz off the perfect fifth. (c) Combining two tones of frequencies 220 and 435 Hz, which is 5 Hz of the octave.

every 3×3.03 ms $= 0.0091$ s, which is close to what we have here. Figure 5.6(b) shows the combined waveform of two tones of equal amplitude, and frequencies 220 and 335 Hz. The frequency ratio here is 335:220 $= 1.523$, which is close to the 1.5 ratio of the perfect fifth, which would correspond to 220 and 330 Hz. So, the 335 Hz is 5 Hz above what the fifth would be. We note here a regular variation in the envelope, that repeats with a period of 0.1 s, which corresponds to 10 Hz, that is, twice the deviation from the perfect fifth. What is interesting here is that this variation in the envelope produces audible beats. In other words, we have audible beats from two frequencies that differ by 115 Hz, which is much higher than the beats discussed in section 1.4. These are called **second order beats**. If the deviation from the fifth was smaller, say 3 Hz, then the frequency of the second order beats would be $2 \times 3 = 6$ Hz[13]. Also, as the deviation from the fifth gets smaller, the magnitude of the variation of the envelope gets smaller.

Figure 5.6(c) shows what happens when we combine two tones with a ratio close to 2:1, that is, close to the octave. In the figure the two frequencies are 220 and 435 Hz. So, the higher frequency is 5 Hz lower than the octave (the octave would be at 440 Hz). Here the envelope varies with a period of 0.2 s, that is, a frequency of 5 Hz, which equals the frequency deviation from the perfect octave. Again, if the deviation from the octave were smaller, say 3 Hz, then the frequency of the second order beats would be 3 Hz as well, and the magnitude of the variation of the envelope would get smaller.

Audio 5.3 captures the second order beat around the fifth. In this recording, one of the tones is 220 Hz, and the second tone sweeps the range of frequencies from about 318–342 Hz in steps of 1 Hz per second. So, the fifth, that is, frequency 330 Hz, is approximately at the midpoint of the recording. The second order beats are noticeable on both sides of the midpoint.

Audio 5.4 captures the second order beat as we approach the octave. In this recording, one of the tones is 220 Hz, and the second tone starts at 410 Hz, and increases up to 440 Hz (that is up to the octave) in steps of 1 Hz per second. What is

Audio 5.3. Second order beats near the fifth of 220 Hz. Available at https://iopscience.iop.org/book/978-0-7503-3539-3.

Audio 5.4. Second order beats as we approach the octave of 220 Hz. Available at https://iopscience.iop.org/book/978-0-7503-3539-3.

---

[13] See explanation in appendix B.3.

interesting here is that the magnitude of the variation diminishes as we approach the octave, but the beats remain very audible, which suggests that the second order beats do not reflect the variation of the envelope alone. The second beats have been known for over a hundred years now, and used by piano tuners to tune octaves, fifths and other intervals.

5.12 Problems and questions

1. (a) In what way does the heptatonic scale differ from the chromatic scale?
    (b) What is the major advantage of the equal temperament scale compared to the Pythagorean and the just scales?

2. (a) Which major scale(s) can be played using only the white keys of the keyboard?
    (b) For each scale of part (a), name the notes corresponding to the major third, the fourth and the fifth.
    (c) Which minor scale(s) can be played using only the white keys on the keyboard?
    (d) For each scale of part (c), name the notes corresponding to the minor third, the fourth and the fifth.

3. Using 440 Hz for A4 is a recent standard. In the Baroque era, A4 = 415 Hz was commonly used.
    (a) Is the present standard sharper or flatter than the baroque standard?
    (b) Find the ratio of the frequencies of A4 (present) and A4 (baroque).
    (c) What is the value of one semitone interval in the equal temperament scale?
    (d) Recalling that the interval between two tones is given by the ratio of their frequencies, how far apart is the Baroque A4 from the modern A4?

4. Use figure 5.1 and the patterns of semitone steps in section 5.2 to construct:
    (a) the A-major scale
    (b) the natural C-minor scale

5. In section 5.8.1 we made seven modes out of a major scale. In our example we started with the C-major scale. Would it be possible to make seven modes starting from a minor scale?

6. Suppose we want to construct a musical scale consisting of six equal intervals. Starting from middle C. So, including the 'octave' we will have seven notes in total. List the notes and frequencies of all the notes in this scale. (*Hint*: How many intervals in the ordinary chromatic scale?)

7. (a) What note is the minor third of A4?
    (b) What is the frequency of the note of part (a) in the Pythagorean tuning?
    (c) What is the frequency of the note of part (a) in the just tuning?
    (d) What is the frequency of the note of part (a) in the equal temperament scale?
    (e) Can you repeat the parts (b)–(d) for the minor fifth of A4? (*Hint*: this may be a trick question).

IOP Publishing

The Physics of Sound Waves (Second Edition)
Music, instruments, and sound equipment
Panos Photinos

Chapter 6

Standing waves and resonance

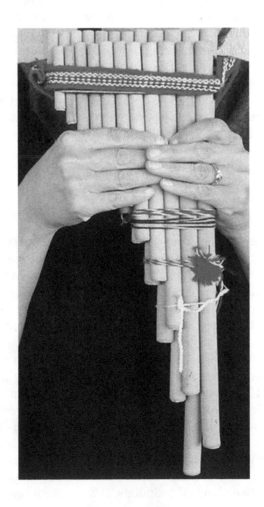

6.1 Introduction

So far, we have described waves that propagate in open space. Propagating waves travel at a speed determined by the medium. For sound waves in air, there are no restrictions on the frequency of the wave, that is, all audible frequencies can propagate in open air. A different type of wave motion occurs when the wave is confined. Take, for instance, the air inside a glass bottle. When blowing air across the mouth of an empty bottle, a characteristic tone is heard. The frequency of the tone produced depends on the size of the bottle. One may ask: why do we hear just one pitch? The tone heard is not produced by vibrations of the glass, but by vibrations of the air *inside* the glass bottle. This is so because blowing air in a half-filled bottle produces a sound of higher pitch. This observation suggests that the frequency of the tone becomes higher as the volume of air inside the bottle gets smaller. In other words, there must be a relation between the tone pitch and the *amount of air in the bottle*. By choosing a set of bottles of the right size and shape, one can create a kitchen version of a pan flute. A pan flute, shown in the opening figure, is essentially a collection of pipes open at one end and closed at the other end. The shape of the tube determines the tone produced by each tube. So, one could use this relation to select the musical tones produced by the instrument. This is the basis of constructing wind instruments, as we will see in section 11.4. A similar relation between length and frequency occurs in strings, and is the basis of string instruments. The vibrational motion excited in pipes and strings is an example of **standing waves**. In this chapter, we discuss the relation between the geometric dimensions and the frequency of the standing waves produced in pipes and strings. The vibrations of plates and membranes will be discussed in sections 11.3.2 and 11.3.3. We will start by discussing standing waves in strings.

6.2 Vibrational modes in a string

Consider a string that is clamped at both ends; for example, a guitar string. The fact that the two endpoints of the string are clamped, means that the displacement is always zero at the endpoints. The string can *only* sustain the waveforms of vibration that are compatible with the requirement that the displacement at the two endpoints remains zero at all times. We will call this condition the **boundary condition**. Plucking a string excites a vibration, in which each point of the string **oscillates** about the equilibrium position. Of course, the two endpoints do not oscillate because they are clamped. In this oscillation, the restoring force (see section 1.2) is the tension of the string. Once plucked, the string quickly settles into a vibration pattern and in many cases, this vibration could produce an audible sound of steady pitch, that could last for several seconds. This is what is meant by 'sustainable vibration' and if we pluck the string again, harder or softer, we will hear sounds of different quality, but of the same pitch. In other words, the string has a preferred pitch. If the string is plucked softly, the pattern produced by the displacement of each oscillating point along the vibrating string follows a sinusoidal wave pattern. This pattern is called a **standing wave**, as opposed to the traveling waves discussed in chapter 2. The pattern is confined between the two clamped endpoints. If the pattern is a single sinusoid, then we can speak of the wavelength of the pattern, as discussed in section 1.2. Note though that this wavelength of the standing wave has nothing to do with the wavelength of the sound produced by the vibrating string, as will be discussed below. But first we will find which standing waves can be sustained by a clamped string. To do so, we will use a simple argument involving the boundary condition[1].

Figure 6.1 shows one of the standing waves on a string. The amplitude of the oscillation is different at each point of the string. At the endpoints the amplitude is zero, because the two ends of the string are clamped. In the middle of the string the amplitude is at maximum. Figure 6.1 shows 'snapshots' of the vibrating string. The sinusoidal pattern drawn corresponds to half a cycle of a sinusoid, that is, 1/2 of a wavelength. This fact allows us to calculate the wavelength of the standing wave. If *half the wavelength* equals the length of the string, then the wavelength is twice the length of the string. So, one standing wave that is allowed is one of wavelength equal to twice the length of the string.

The pattern shown in figure 6.1 is not the only sinusoid compatible with the boundary condition. As it turns out, any sinusoid that satisfies the boundary condition represents a vibration that can be sustained by the string. Figure 6.2 shows another sinusoid compatible with the boundary condition, in other words, a vibration that can be sustained by the string. In this case, as indicated by the green line in the figure, one full cycle (or two half-wavelengths) of the sinusoid equals the length of the string.

Length of string = 1 λ /2

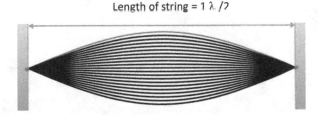

Figure 6.1. Snapshots of a standing wave pattern in a guitar string.

Length of string = 2 λ /2

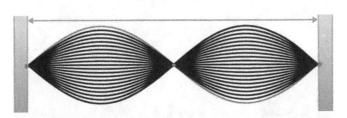

Figure 6.2. A standing wave of two half-cycles.

---

[1] A mathematical derivation is given in appendix B.4.

Figure 6.3 shows yet another waveform compatible with the boundary condition. As indicated by the green line in the figure, the sinusoid completes one and one-half cycles, or three half-cycles. This tells us that the length of the string equals 3/2 wavelengths; in other words, the wavelength is 2/3 of the length of the string.

Videos 6.1(a–c) show the vibration of the string for the first three modes.

Comparing figures 6.1–6.3, we see that the wave in figure 6.1 has the longest wavelength of all. This is the first or **fundamental mode** of the vibration. The next

Length of string = 3 λ /2

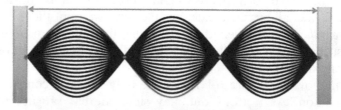

Figure 6.3. A standing wave of three half-cycles.

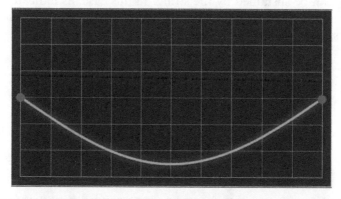

Figure Video 6.1(a). First mode. Video available at https://iopscience.iop.org/book/978-0-7503-3539-3.

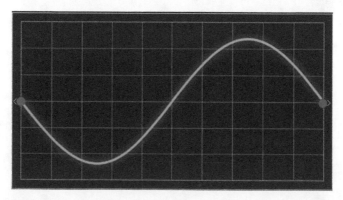

Figure Video 6.1(b). Second mode. Video available at https://iopscience.iop.org/book/978-0-7503-3539-3.

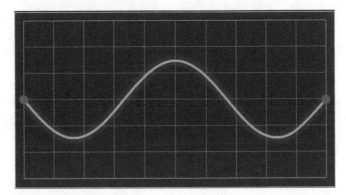

Figure Video 6.1(c). Third mode. Video available at https://iopscience.iop.org/book/978-0-7503-3539-3.

longest wavelength occurs in the mode shown in figure 6.2. This is the second vibrational **mode.** In figure 6.3, the wavelength is shorter than the previous two. This is the third vibrational mode. It follows that a sinusoidal vibration is allowed (in other words it satisfies the boundary condition) as long as we can fit an integer number of half-wavelengths into the length of the string. So, a vibration is sustained whenever:

$$n \times (\text{wavelength}/2) = \text{length of string}$$

where n can be 1, 2, 3, 4, on so forth. For $n = 1$ we have the fundamental mode, $n = 2$ is the second mode, and so on. Comparing the wavelengths from figures 6.1–6.3 we see that there is a direct relation between the wavelength and the order of the vibrational mode:

- The wavelength of the second mode is 1/2 wavelength of the fundamental.
- The wavelength of the third mode is 1/3 wavelength of the fundamental.

By adding half-wavelengths as done above, one can see a trend in how the wavelengths relate to the fundamental:

- The wavelength of the fourth mode is 1/4 wavelength of the fundamental.
- The wavelength of the nth mode is $1/n$ the wavelength of the fundamental, where n is any positive integer (1, 2, 3, and so on).

Example: As an application we consider the vibration modes of the fifth string (fifth from bottom) of a guitar. In standard tuning, the open fifth string produces a fundamental mode of frequency equal to 110 Hz. When the string is open, the two endpoints[2] are the **saddle** (on the body of the guitar) and the **nut** (at the top of the fretboard). Typically, the distance between saddle and nut is about 0.65 m.

As discussed above, the length of the string is equal to 1/2 wavelength of the fundamental mode, so, the wavelength of the fundamental mode is $2 \times 0.65 = 1.3$ m. It follows that:

- The wavelength of the second vibrational mode equals 1.3/2 = 0.65 m.
- The wavelength of the third vibrational mode equals 1.3/3 = 0.433 m, and so on.

---

[2] See section 11.5.1 for the parts of a guitar.

Recall from section 3.5 that the guitar string produces a number of overtones (see figure 3.11) with frequencies that, *within the accuracy of our measurements*, are multiples of the fundamental. In this case, we expect to have frequencies that are multiples of 110 Hz, that is 220, 330, 440 Hz, and so forth, each corresponding to a vibrational mode. So, we have the following pattern for the mode frequencies: 1, 2, 3, 4, 5, ... times the fundamental. This pattern is the **harmonic series**. This is not always the case. As we shall see below, we could have other patterns, for instance, it is very common to have mode frequencies that are 1, 3, 5, 7... times the fundamental, that is odd multiples of the fundamental.

Table 6.1 lists the wavelengths of the standing waves and frequencies of the first four modes for the fifth string of a guitar. The length is assumed to be 0.65 m.

Table 6.1. Wavelengths and frequencies of the vibrational modes of the fifth string of a guitar.

Mode	Mode wavelength (m)	Mode frequency (Hz)
$n = 1$	1.30	110
$n = 2$	0.65	220
$n = 3$	0.433	330
$n = 4$	0.33	440

Recall from section 1.2 that the frequency of the wave is related to the wavelength and the speed of the wave by the equation:

$$\text{speed} = (\text{frequency}) \times (\text{wavelength})$$

We can apply this equation for each mode listed in table 6.1 and find the speed of the wave in the string. Doing so we have:

Fundamental mode:	110 Hz × 1.3 m = 143 m s^{-1}
2nd vibrational mode:	220 Hz × 0.65 m = 143 m s^{-1}
3rd vibrational mode:	330 Hz × 0.433 m = 143 m s^{-1} and so on.

Note that the speed is the same for all vibrational modes. The speed of the wave in the string depends on the magnitude of the restoring force (that is the tension of the string) and the nature of the string (specifically the thickness and the material). It follows then that, for a given string, the *speed is the same for all modes*. Also note that the speed of 143 m s^{-1} refers to the speed of the waves in the string, *not* the speed of sound in air, which is about 343 m s^{-1}. We also see that the wavelengths listed in table 6.1 refer to the vibration modes *of the string* and *not* to the wavelength of the sound produced by the string. To find the wavelength of the resulting sound wave in air we divide the speed of sound in air by the frequency of each mode. Doing so we find 343/110 = 3.12 m for the fundamental, 343/220 = 1.56 m for the second mode, and so on.

One may wonder what is the meaning of speed for a standing wave? After all, the standing wave is not going anywhere! For one, the speed mentioned here is *not* the speed at which the string is moving up and down. Actually, different parts of the string move at different speeds! To understand what this speed of the wave in the string represents, consider a very long string clamped at both ends. If we pluck the string at one end, that would cause a disturbance that travels towards the other end. If we divide the length of the string by the time it takes for the disturbance to travel the entire length of the string, we get the speed of the wave *in the string*[3]. This is what the value of 143 m s^{-1} in the above list refers to.

6.3 Nodes and antinodes

From figure 6.1 we note that the fundamental mode of vibration has maximum amplitude at one point (in the middle) and zero amplitude at two points (the endpoints). In the second mode (figure 6.2) the amplitude is maximized at two points, and is zero at three points. Similarly, for the third mode (figure 6.3) the amplitude has three maxima, and four zeros. Following this trend, the fourth mode will have four maxima and five zeros, and so on. Points on the string that do not oscillate (that is, where the amplitude is zero) are referred to as **nodes**. Points where the oscillation is maximized are **antinodes**.

6.4 Simultaneously vibrating modes

As we know from section 3.5, guitar strings oscillate in many frequencies at the same time. This means that the tone produced by plucking a string is *not* a **pure tone** (that is *not* a single frequency) but a **complex tone**, consisting of many frequencies or **partials.** The term *partial* is used to indicate any of the vibrating modes, that is the fundamental plus the overtones. In terms of standing waves, each partial is associated with a vibrational mode, and the presence of multiple modes vibrating simultaneously plays a key role in what we call quality or **timbre** of a tone, discussed in section 3.6. What this means is that the modes shown in figures 6.1–6.3 almost never occur separately. Strings in actual instruments vibrate in a much more complicated fashion. Examples of the vibration of a string when more than one mode is present are shown in videos 6.2(a)–(c). In video 6.2(a) we have modes 1 and 2. The amplitude of mode 1 was chosen to be 4 times larger than the amplitude of mode 2. In video 6.2(b) we have modes 1 and 3. The amplitude of mode 1 was chosen to be 4 times larger than the amplitude of mode 3. In video 6.2(c) we have modes 1, 2 and 3. The amplitudes of all three modes are chosen to be equal.

Which harmonics oscillate and how strongly depends to some extent on *where* and *how* the string is excited to vibration, and musicians can to some degree control the modes that oscillate, in other words, they can control the timbre of the tone. One simple method suggests itself by examining figure 6.2, which shows that the second vibrational mode has a node at the midpoint. So, by plucking a guitar string at the midpoint, we force motion at the midpoint, and that is not compatible with the

---

[3] This speed of the up-and down motion of the string is discussed in appendix B.4.

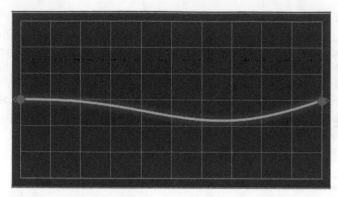

Video 6.2(a). Mode 1 and mode 2. Video available at https://iopscience.iop.org/book/978-0-7503-3539-3.

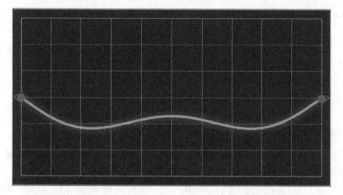

Figure Video 6.2(b). Mode 1 and mode 3. Video available at https://iopscience.iop.org/book/978-0-7503-3539-3.

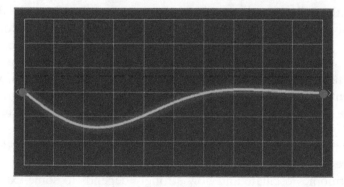

Video 6.2(c). Modes 1, 2 and 3. Video available at https://iopscience.iop.org/book/978-0-7503-3539-3.

second mode, because the second mode must have a node in the middle. In other words, *plucking the string at the midpoint* will suppress the second harmonic. Note that all even modes (like the fourth, sixth, and so on) have an odd number of nodes, meaning that the midpoint for all even harmonics must be a node. It follows then that by plucking the string at the middle, we *suppress all the even modes*.

6.5 Standing waves in pipes

Standing waves occur in pipes as well, and this is the basis of wind instruments. Here the vibration of interest is a longitudinal vibration of the air column *inside* the pipe, not the vibration of the pipe walls. As discussed in section 1.2 the wave consists of compressions and rarefactions, meaning that the pressure in the air column gets high and low *compared to the ambient* (or *atmospheric*) pressure. The difference between the total pressure inside the pipe and the ambient pressure is what we called the *acoustic pressure* in section 2.3. Here we will describe the standing waves in a pipe using the concept of nodes and antinodes introduced in section 6.3. For musical instruments we have two types of pipes that are of interest, namely: a pipe open at both ends, and a pipe that is open at one end and closed at the other.

As in the case of standing waves in a string, the starting point is the boundary condition. For a clamped string, the two endpoints do not oscillate, so, we have a node at both ends. In the case of a pipe, we have two possibilities for each end: *open* or *closed*, and each imposes a different boundary condition:

At the open end of a pipe, the total pressure must be equal to the ambient pressure at all times. This means that the acoustic pressure, which is the difference between total pressure and ambient pressure must be zero at all times at the open end of a pipe. So, **at the open end of a pipe the acoustic pressure must have a node.**

At the closed end of a pipe, the air can be compressed to a maximum pressure and then bounce back to a minimum pressure. This means that at the closed end the changes in acoustic pressure are the largest. The conclusion is that **at the closed end of a pipe the acoustic pressure must have an antinode.**

We will use the above boundary conditions to investigate the mode frequencies of the standing waves supported by two types of pipes.

6.6 Standing waves in an open pipe

For the open pipe, the boundary conditions are the same as the clamped string, that is, **the two endpoints are pressure nodes.** Using the same arguments as in section 6.2 we conclude that the open pipe will sustain standing waves if the length of the pipe is equal to an integer number of half-wavelengths.

Modes in open pipe:

$$n \times (\text{wavelength}/2) = \text{length of pipe}$$

where n can be 1, 2, 3, 4, etc. For $n = 1$ we have the fundamental mode, $n = 2$ is the second mode, and so on.

Figure 6.4 shows the fundamental mode in an open pipe which is identical to figure 6.1. What is different is the nature of the oscillating quantity. In the case of the string (figure 6.1) the graphs show snapshots of the displacement of the string from the equilibrium position. Figure 6.4 shows snapshots of the variation of the acoustic pressure in the pipe. The acoustic pressure at the two ends of the pipe is zero. With this understanding, one can interpret figures 6.2 and 6.3 as showing the second and third harmonics in an open pipe.

Length of Pipe

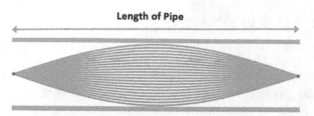

Figure 6.4. Fundamental mode in an open pipe.

We can rearrange the quantities in the above relation to find the wavelengths of the modes in an open pipe:

Mode wavelengths in open pipe:

$$\text{wavelength} = (2/n) \times \text{length of pipe}$$

where $n = 1, 2, 3, 4$, and so on. From the above relation we conclude that longer pipes produce fundamentals of longer wavelengths.

Example: For an open pipe of length 1.70 m, find the wavelength of the fundamental mode.

For the fundamental mode $n = 1$, applying the mode wavelength relation we have:

$$\text{wavelength} = (2/1) \times 1.70 = 2 \times 1.70 = 3.40 \text{ m}.$$

Since the product: frequency × wavelength equals the speed of the wave, it follows that the frequency is equal to the ratio of the speed divided by the wavelength. Here the speed of the wave is the speed of sound in air, that is 343 m s^{-1}. Using the mode wavelengths from the above relation, we can find the mode frequencies (in Hz) for the open pipe:

Mode frequencies in open pipe:

$$\text{frequency} = (n \times 343)/(2 \times \text{length of pipe}).$$

In the above relation, the length of the pipe is in the denominator, which means that the longer the pipe, the lower the frequency of the fundamental.

Example: For an open pipe of length 1.70 m, find the frequency of the fundamental mode.

For the fundamental mode $n = 1$, and applying the mode frequency relation, we have:

$$\text{frequency} = (1 \times 343)/(2 \times 1.70) = 343/3.4 = 101 \text{ Hz}.$$

Table 6.2 lists the wavelengths and frequencies for the first four modes in an open pipe of length 1.70 m. We note that the frequency of the modes follows the harmonic series, which was introduced in section 6.2.

Table 6.2. Wavelengths and frequencies for the first four modes in a 1.7 m long open pipe.

Mode	Wavelength (m)	Frequency (Hz)
$n = 1$	3.40	101
$n = 2$	1.70	202
$n = 3$	1.13	303
$n = 4$	0.85	404

6.7 Standing waves in a pipe closed at one end

Pipes with one end open and one end closed are also referred to as **stopped pipes**, or simply **closed** pipes. In a closed pipe, we must have a **pressure node at the open end** and a **pressure antinode at the closed end**. A half-wavelength has nodes at both ends, so it is not compatible with boundary conditions. To have an antinode at the closed end we need a quarter-wavelength.

Length of pipe

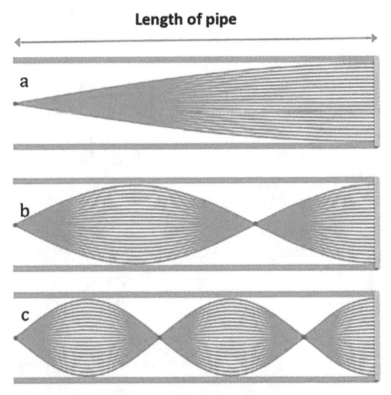

Figure 6.5. First three modes for a closed pipe.

Figure 6.5 shows the first three modes for a closed pipe. For the fundamental mode, shown in figure 6.5(a), we note that the length of the pipe is equal to 1/4 of the wavelength. For the next higher mode, shown in figure 6.5(b), the length of the pipe is equal to 3/4 of the wavelength. The length of the pipe in figure 6.5(c) is equal to 5/4 of the wavelength. Comparing figure 6.5(a)–(c) we conclude that successive modes differ by 1/2 wavelength, or two quarter-wavelengths. Note that the fundamental fits one quarter-wavelength in the length of the pipe. So, the sequence of modes in terms of *quarter-wavelengths* will be (0 + 1), (2 + 1), (2 + 2 + 1), (2 + 2 + 2 + 1), and so forth, or 1, 3, 5, 7, … *quarter-wavelengths*.

Following this trend, we can write:
Modes in closed pipe:

$$n \times (\text{wavelength}/4) = \text{length of pipe}$$

where n follows the sequence of odd numbers, that is, 1, 3, 5, 7, and so on. For $n = 1$ we have the fundamental mode, $n = 3$ is the second mode, $n = 5$ is the third mode, and so on.

Rearranging the quantities in the above relation we can find the wavelengths of the modes in a closed pipe:

Mode wavelengths in closed pipe:

$$\text{wavelength} = (4/n) \times \text{length of pipe}$$

where $n = 1, 3, 5, 7$, and so on.

Example: Find the wavelength of the fundamental mode in a closed pipe of length equal 0.85 m.

For the fundamental mode $n = 1$, applying the mode wavelength relation we have:

$$\text{wavelength} = (4/1) \times 0.85 = 4 \times 0.85 = 3.40 \text{ m.}$$

Following the procedure used in connection with open pipes in section 6.6, we can find the mode frequencies (in Hz) for the closed pipe:

Mode frequencies in closed pipe:

$$\text{frequency} = (n \times 343)/(4 \times \text{length of pipe}).$$

Example: For a closed pipe of length 0.85 m find the frequency of the fundamental mode.

For the fundamental mode $n = 1$. Applying the mode frequency relation, we have:

$$\text{frequency} = (1 \times 343)/(4 \times 0.85) = 343/3.40 = 101 \text{ Hz.}$$

Table 6.3 lists the wavelengths and frequencies for the first four modes in a closed pipe of length 0.85 m.

6.8 Comparison of open and closed pipes

Open and closed pipes share some common properties. The longer the pipe, the longer the wavelength of the modes, and the lower their frequencies. The fundamental

Table 6.3. Wavelengths and frequencies for the first four modes in a 0.85 m long closed pipe.

Mode	Wavelength (m)	Frequency (Hz)
$n = 1$	3.40	101
$n = 3$	1.13	303
$n = 5$	0.68	504
$n = 7$	0.49	706

frequency is higher for shorter pipes. But there are also two essential differences between the two types of pipes, which can be seen by comparing the values listed in tables 6.2 and 6.3. The first difference is that the closed pipe produces only odd multiples of the fundamental frequency, while the open pipe produces both odd and even multiples of the fundamental frequency. This means that the timbre of the two types of pipes is different. Second, we note that the closed pipe (0.85 m long in the above example) produces the same fundamental frequency as an open pipe twice as long (1.70 m in the example). It follows that closed wind instruments (see section 11.4) can be shorter than open instruments, and produce the same range of tone frequencies.

6.9 Standing waves in a pipe closed at both ends

Finally, applying the boundary conditions to a pipe closed at both ends, we see that the acoustic pressure must have an antinode at both ends, as shown in figure 6.6. In the first mode we have half a wavelength; the second mode has a full wavelength, and so on. It is not difficult to see that the mode wavelengths are given by:

$$\text{wavelength} = (2/n) \times \text{length of pipe}$$

and the **mode frequencies in closed pipe** are given by:

$$\text{frequency} = (n \times 343)/(2 \times \text{length of pipe})$$

where n takes the values 1, 2, 3, and so on. In other words, the equations for the wavelengths and frequencies of the modes are the same equations as in the case of the open pipe.

$$\text{frequency} = (n \times 343)/(2 \times \text{length of pipe})$$

Figure 6.6. Fundamental mode in a pipe closed at both ends.

Pipes closed at both ends are not relevant to musical instruments, since the sound cannot escape from the closed pipe. However, the pipe closed at both ends is a useful model for standing waves in closed spaces, as will be discussed in section 9.11.3.

6.10 Standing waves in rods and tubes

Of interest to percussion instruments are standing waves that form in solid rods and in tubes. Solids can transmit both longitudinal *and* transverse waves, and can sustain both longitudinal and transverse standing waves. The behavior of rods and tubes is essentially the same, and we will use *rods* to refer to both. The frequency of the fundamental mode depends on the length of the rod. The frequency of the fundamental is lower for longer rods. This trend is similar to what we found for strings and air columns in pipes. The frequency of the fundamental mode depends also on the stiffness of the material. For example, the fundamental frequencies of wooden or bamboo rods are lower than those of steel or aluminum. The stiffness of the material acts as the restoring force (see section 1.2) so it is reasonable that wood, which bends more easily than steel, will have a weaker restoring force, and bounce back more slowly than steel. In the case of strings, the restoring force is set by the tension applied on the string, which means that the higher the tension, the higher the frequency of the fundamental. The difference is that in musical instruments, the rods are not under any tension. Here we will discuss a case of interest to percussion instruments, namely a rod free at both ends.

Figure 6.7(a) shows the fundamental mode for a rod that is free at both ends. This can be realized, for example, by suspending the rod from a string, which is the basic principle of chimes, including common wind chimes. We note that the fundamental mode has two nodes, the second mode will have three nodes, and so on. The frequency of the second mode is not an integer multiple of the frequency of the fundamental. Instead, it is 2.78 times the frequency of the fundamental. The amplitude of the second mode may be high, even higher than the amplitude of the fundamental, if the rod is struck midway between the center and one of the free ends. The tuning fork, shown in figure 6.7(b) is a variation of the rod free at both ends. Essentially, a tuning fork is a bent rod so the fundamental mode has two nodes, as in figure 6.7(a) located near the bottom of each prong. Tuning forks were widely used in the past for tuning musical instruments. Note that the midpoint between the two prongs is not a node. It vibrates in the direction of the stem, meaning that there is a longitudinal vibration in the stem. This is the vibration involved in sound healing with tuning forks (section 4.13.5.)

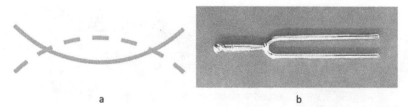

a   b

Figure 6.7. (a) Fundamental mode of a rod free at both ends. (b) Tuning fork.

6.11 Harmonics, partials, and overtones

As the various modes oscillate simultaneously, the musical tones produced by strings, pipes, rods, and the like, are complex tones, in that they contain many frequencies. An **overtone** is any mode above the fundamental. For instance, the second mode in a string is the first overtone. **Partial** is any mode, in other words, the fundamental or an overtone. So, the fundamental is the first partial, the second mode is the second partial, and so on. If the overtones have frequencies that are integer multiples of the frequency of the fundamental, then we speak of **harmonic overtones**, or simply **harmonics**. Such was the case with the mode frequencies of the vibrating string, and the air column in pipes. If the overtone frequencies are not an integer multiple of the frequency of the fundamental mode, then we have **non-harmonic** or **anharmonic overtones**. Such was the case of the rods discussed in the previous section. Anharmonic overtones occur also in plates and stretched membranes, as will be discussed in section 11.3.

6.12 Resonance and damping

An oscillator is a system whose state can change over periodic cycles; for example, a swing, a string, and the like. If a swing is given one push, or a string is plucked, then what follows is a free oscillation. A free oscillator has its characteristic or **natural frequency** (or frequencies). The natural frequency of a swing is determined by the length of the swing. In a similar way, a string has a number of natural frequencies, its vibrational modes, each with its own frequency, as discussed in section 6.2. For simplicity we will use the swing as our example in what follows. When the swing is pushed again and again, the swing becomes a **driven oscillator**. Children discover early-on that they can drive the oscillator themselves by *pumping*: leaning back and extending the knees when the swing is moving forward, then bending the knees and leaning forward when the swing is moving back. Either way, pushing or pumping at random does not necessarily make the amplitude of the oscillation larger. This can only be achieved if the swing is driven with proper timing, which is different for each swing.

As it turns out, if the frequency of the driving force is low compared to the natural frequency of the swing, the amplitude of the oscillation will be small. The same happens if the frequency of the driving force is high compared to the natural frequency of the swing. If the frequency of the driving force is close to the natural frequency of the swing, then the amplitude of the oscillation can become very large. In this case we have **resonance**. Exactly how large the amplitude can get depends on the energy loss associated with the motion of the swing. For instance, if the swing is too low and the child's feet are dragging on the ground on every cycle, the amplitude at resonance will be smaller. For an oscillator, this energy loss could arise from friction, air resistance, or other factors, which we will refer to as **damping**. How close the driving frequency has to be to get the maximum amplitude depends on the amount of damping present.

Figure 6.8(a) shows the amplitude of the oscillation for three different cases. The blue line shows the amplitude of the oscillator with very little damping. The red line

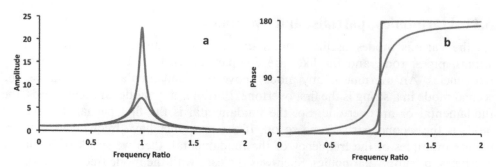

Figure 6.8. (a) Amplitude of a driven oscillator for three values of damping. The peak is narrow and high when damping is low. (b) Phase difference, in degrees, between driving force and displacement.

shows the amplitude when damping is increased ten times, and the green line shows the amplitude when damping is increased 100 times. The horizontal axis is the ratio of frequency of the driving force divided by the natural frequency of the oscillator. So, at the value '1' the oscillator is driven at its *natural frequency*. The driving frequency is lower than the natural frequency for values less than one and vice versa. The vertical axis shows the amplitude of the oscillation in arbitrary units. For the case of low damping (blue line) the peak value corresponds to an over 20-fold increase, compared to the values at the left-end of the graph. The peak occurs when the driving frequency is equal to the natural frequency. At this point we have resonance and the amplitude of the oscillation assumes a maximum value. For the intermediate case (red line) the peak is lower and the curve is broader, and we have a resonance that is not as sharp or strong. For high damping (green line) the peak is very broad and very low. Note also that the resonance occurs at a driving frequency slightly lower than the natural frequency of the oscillator.

Figure 6.8(b) shows the phase difference between the driving force and the displacement. We see that if the frequency of the driving force is too low compared to the natural frequency of the oscillator, then the phase difference is zero so the force and the displacement are in-step. In the case of the swing, this would be like pushing forward when the swing is already at its most forward point of the cycle. We also see from the figure that when the frequency of the driving force is too high compared to the natural frequency of the oscillator, then the force and the displacement are 180° out-of-step, and this would be like pulling the swing backward while it is trying to reach its most forward point. In both cases, the result is that the amplitude of the oscillation is small. When the frequency of the driving force is equal to the natural frequency of the oscillator, the force is 90° ahead of the displacement. This means that when the swing is at the lowest point moving forward, then the force is in the forward direction and at its maximum. When the swing reaches the maximum displacement (forward or backward) the force is zero, and starts building up gradually.

6.13 Examples of resonance

One can use a guitar to observe resonance in strings. The effect is more obvious if one uses the two top strings. If the guitar is tuned, pressing one finger at the fifth fret of the top (E) string corresponds to the same tone as the open fifth string (A). When the top string is plucked, the A string will begin to vibrate as well. The vibrations of the top string are transmitted through the saddle of the guitar (see figure 11.13) to the A string, and since the frequencies match, we have resonance. If the finger is at the sixth fret, plucking the top string has no effect on the A string, because the frequencies are not the same, therefore, there is no resonance.

Some instruments take advantage of this resonance between strings. For example, the viola d'amore and the Indian sitar, among others, have a set of 'sympathetic' strings[4]. The musician does not play these strings. Instead, the sympathetic strings are tuned to resonate with some of the tones that are produced by the played strings, which makes the tone last a bit longer, and produce a fuller sound.

The vibrational energy of a string does not transfer efficiently to the surrounding air. One way[5] to overcome this is to transferred the vibrational energy to a box (the body of a violin, for instance) and have the sound energy radiate from the box to the surrounding air, which is much more efficient. For this transfer to happen, the box must have resonances that match the frequencies of the vibrating strings. These resonances should be sharp so that the resulting sound is loud. On the other hand, these resonances should occur at frequencies that cover the entire tonal range of the instrument. For most string instruments this range is usually over three octaves. In other words, one should have a huge number of resonances, which is not practical. In practice the instrument makers choose a small number of overlapping wide resonances to cover the entire tonal range. This is basically what happens in most string instruments, as discussed in section 11.5.

Because of the associated energy loss, one may be tempted to think that damping is an undesirable quantity. This is not always the case. In other words, there are applications where a sharp resonance (blue line in figure 6.7) is desirable, and other applications where a broad resonance (red line in figure 6.7) or very low resonance (green line in figure 6.7) are the most desirable. This will be illustrated by the examples that follow.

As a rule, sharp resonances should be avoided in most structures and mechanical systems, because the amplitude of the oscillation at resonance can get high enough to degrade the sound quality or even damage the oscillator[6]. In an indoor listening environment, all resonances in the audio frequency range arising from walls, windowpanes, furniture, loudspeaker enclosures and the like, must be over-damped to extinction. Closed spaces behave like pipes closed at both ends (see section 6.9) in that they can sustain a set of vibrational modes. As a result, some frequencies may

---

[4] The term 'sympathetic vibration' is used to indicate transfer of energy from one oscillator to another. This transfer of energy becomes most efficient when the natural frequencies of the two oscillators are close. So, sympathetic vibrations are driven vibrations.

[5] Another way is to use an amplifier, as is done in electric guitars. Note that if the amplifier of an electric guitar is off, the sound from the strings is very weak.

[6] There are accounts of breaking wine glasses with one's voice. The vibrational modes of typical wine glasses are within the range of the human voice, and intensity levels over 100 dB (see section 4.4) at the resonance frequency of the glass can shatter the glass, according to these accounts. The author has not witnessed such a formidable event.

sound louder than intended. This problem is particularly noticeable with car sound systems. Let's assume that the width of a car's interior is around 1.5 m (about 5 ft). Then the mode equation from section 6.9 tells us that the frequency of the fundamental (figure 6.6) is $343/(2 \times 1.5) = 114$ Hz, which is in the bass range (see section 4.2). The resonance can be high enough to make the side windows of the car move visibly, and that should tell us that this resonance is clearly undesirable. Fortunately, the solution to this problem is very simple!

Another example of undesired sharp resonance is the cone of a loudspeaker (see section 7.7) which is a mechanical oscillator driven by the output of the amplifier. Without damping, the cone would respond strongly to a certain frequency and much less to the other sound frequencies, which would degrade the quality of sound. In addition, the high amplitude of oscillation at resonance may be enough to destroy the cone.

Sharp resonance is most important in radio wave communication, which includes, radio, TV, WiFi, and the like. Each location receives all sort of signals from various sources at once. So, it is important for the receiver to pick up the desired carrier frequency, and ignore all other frequencies. This is a most critical step, otherwise, on a radio tuner for example, we would be listening to all the stations at the same time. Tuning to just one station is accomplished by running all the signals that reach the location of the receiver, through an electrical oscillator circuit that has very sharp resonance, in other words, an oscillator with very little damping.

6.14 Further discussion

6.14.1 Resonance in air cavities

In section 6.1 we mentioned the familiar example of producing sound by blowing across the mouth of a bottle. In this case we have air in a cavity, and that is slightly different from a straight pipe. As we blow across the mouth of the bottle, the air stream compresses the air in the bottle. At some point, the air pressure at the mouth of the bottle gets sufficiently high, to overcome the compression from the air stream, and begins to push air out of the bottle. Once the pressure in the bottle is released, compression from the air stream takes over, and a new cycle begins. This cycle of compressions and expansions of the air in the bottle creates a series of compressions and rarefactions that propagate outward, in other words, a sound wave. The frequency of the sound produced by this process depends on how long it takes for the pressure to build up in the bottle, and how long it takes the compressed air in the bottle to leak out through the neck of the bottle. It takes longer for the pressure to build up in a large bottle, and also it takes longer for the air to leak out of the bottle if the neck is long or narrow. This tells us that the frequency will be lower if the volume of the bottle is large, or the neck is long or narrow. Conversely, the frequency is higher if the volume of the bottle is small, or the neck is short, or wide. In other words, the bottle has its own natural frequency, just like the swing discussed in section 6.12. The air stream, which acts as the driving force, is essentially noise (see section 5.10) which contains all frequencies, including the frequency that matches the natural frequency of the bottle. So, we can say that from all the frequencies contained in the air stream, the bottle picks up and 'amplifies' the sound frequency that matches its own natural frequency.

Note that the air pressure inside the bottle oscillates at the natural frequency of the bottle, and that it is not necessary to blow across the opening to excite the oscillation. One could just bring any sound source close to the mouth of the bottle. If one of the components of the frequency spectrum of the sound source matches the natural frequency of the bottle, that frequency will be amplified. Moreover, one can use a series of resonators to identify the frequency components of the sound, a Fourier analysis of sorts (see section 3.8) without using electronic equipment! This was done by Helmholtz[7] who used especially designed spherical cavities, with a small nipple that fit into one's ear, so one could hear directly the oscillation of the air inside the cavity.

The sound boxes of ordinary guitars and violins serve as resonators, and as such they have resonance frequencies. Here there is an added complication that the walls of the box are made of thin flexible wood, so the walls oscillate as well. In any case the resonance frequency of the air in the box is typically about 100 Hz for a guitar, and about 200 Hz for a violin. These frequencies are at the low end of the tonal range of the guitar and violin[8]. In other words, this is the resonance that amplifies the low frequencies, as can be demonstrated by noticing the change caused by covering the sound hole. This is done in Video 6.3. In the video, the sound hole of the string instrument (a *bouzouki* in this case) is covered. Once the cover is retracted, the lower frequencies became more audible, because the lower frequencies resonate with the box.

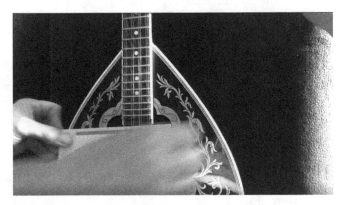

Figure Video 6.3. Lower frequencies resonate in sound box of a string instrument. Video available at https://iopscience.iop.org/book/978-0-7503-3539-3.

6.14.2 Real strings

In our discussion of vibrating strings, we assumed that the tension of the vibrating string is the same as when the string is at rest. This is approximately true. The bends

---

[7] Hermann von Helmholtz, 1821–1894.
[8] See section 11.5.

in the vibrating string make the string longer, which implies that the tension of the vibrating string and the frequency will increase. This dependence of the frequency on the amplitude of vibration can be noticeable, and some artists do it to enrich the sound. But as long as the amplitude of the vibrations at each point of the string is very small compared to the vibrating length of the string, the assumption that the length, and tension remain the same is a good approximation.

We also assumed that the string is perfectly flexible. This can be a weak assumption and, in some cases, has significant practical implications. Strings are not perfectly flexible, and this is more noticeable for the thicker ones. Thick strings are necessary for producing low frequency tones in musical instruments, like pianos, guitars, and the like. To reduce the effects of stiffness, the strings use wound wire around a thin metallic or nylon core. The wire winding adds thickness to the string and at the same time keeps it flexible. But even with winding, the strings are far from perfectly flexible: a bent string will always try to straighten up, which means that there is an additional restoring force, besides the tension of the string. Recall from section 1.2 that a larger restoring force means higher frequency (see also equations in section 6.15). In other words, the frequency of all vibrational modes will be slightly higher than predicted by our analysis in section 6.2. Now if the increase was in proportion to the mode number, that would not be a serious problem. All one would have to do to fix it would be to lower the tension of the string. The problem is that the increase is higher for the higher vibrational modes. We can understand this by looking at figures 6.1–6.3. The fundamental bends the string once, the second vibrational mode bends the string twice, the third bends the string three times and so forth. This means that the overall bending of the string, and the frequency shift, increase with mode number.

To illustrate the point, consider the note A2. Neglecting the stiffness of the string, the three harmonic overtones are multiples of 110 Hz:

220 330 440 Hz and so on.

For argument's sake, let us assume that bending causes the first overtone to increase by 1 Hz, and the increase in the second, and third overtones is 2.2 and 4 Hz, respectively. So, striking the A2 note on the keyboard will produce a tone with the following frequency components:

110 Hz (fundamental);
221 Hz (1st overtone);
332.2 Hz (2nd overtone)
444 Hz (3rd overtone)

We see that the overtones of 110 Hz for a thick string are not multiples of 110 Hz, that is, the overtones are not harmonic. Now consider what happens when we strike A2 and A3 simultaneously. The fundamental of A3 is 220 Hz which clashes with the 1st overtone of A2 which is 1 Hz higher. As discussed in section 1.4 the result would be a sound of frequency of $(221-220)/2 = 220.5$ Hz, and *beat* frequency 1 Hz. The same will happen with the third overtone (444 Hz) and A4, which has frequency 440 Hz. The result would be a sound of 442 Hz beating 4 times per second. So here is the problem: if we tune A2 to 110 Hz, its overtones will produce unpleasant sounding beats with all the A notes. So, what is the solution? In tuning the lower notes of a piano, one has to compromise, usually by tuning, for example the A2 note slightly flat (lower frequency) to minimize clashing with all the fundamentals and overtones of all A notes of higher frequencies, like A3, A4, and so on.

6.14.3 Real pipes

The results derived in sections 6.5–6.7 are exact if the pipes are infinitely long. In practice our results are still valid as long as the diameter of the pipe remains small compared to the length of the pipe, with a slight modification at the open end. Recall that at the open end the acoustic pressure has a node, in other words the total pressure at the open end of the pipe is equal to the atmospheric pressure. For real pipes the node in the acoustic pressure is a bit further out from the open end, by about 0.6 r, where r is the radius of the pipe. This is a correction that needs to be applied to the length of the pipe. For example, in section 6.6 we considered pipes open at both ends. We can use the same equations, but instead of the actual pipe length we must use the corrected length. The correction must be applied to both ends. For example, if the length is 0.4 m and the radius is 0.02 m, then the correction for each end is 0.6 ×0.02= 0.012 m. This means that the effective or 'acoustic' length of the pipe is now:

$$\text{Acoustic Length} = 0.4 + 0.012 + 0.012 = 0.424 \text{ m}.$$

In section 6.7 we discussed pipes open in one end, so the correction applies only once. For example, if the length is 0.4 m and the radius is 0.02 m, the acoustic length is:

$$\text{Acoustic Length} = 0.4 + 0.012 = 0.412 \text{ m}.$$

6.14.4 Other vibrational modes in strings

For strings, the transverse standing waves discussed in this chapter are not the only possible modes of vibration. Strings are solids, and from section 1.2 we know that solids can transmit both transverse and *longitudinal* waves. The restoring force for these longitudinal waves is not the tension of the string, but the elastic response of the material itself. The result is that the frequency of the longitudinal modes in the string is independent of the tension of the string. Also recall from table 2.1 that speed of sound waves in a solid is fairly high compared to the speed of the vibrational modes discussed in section 6.2, meaning that we should expect the frequencies of the longitudinal modes to be high. We can hear the longitudinal modes in a guitar, when sliding a finger along a metal wound string. This is not uncommon, and at times the added sound can be pleasing. We can also hear the longitudinal modes in a violin. The longitudinal modes are excited when the bow is meeting the string at an angle. The result is an unpleasant screech[9].

---

[9] Strings can also support *torsional* modes, where the string essentially twists back and forth, and the resulting sound is generally unpleasant.

6.15 Equations

Speed of wave in a clamped string

The speed c of a wave in a string is given in terms of the tension of the string F and its linear density d (which is the mass of the string divided by the length of the string) by

$$c = \sqrt{F/d}.$$

Mode frequencies in a clamped string

$$f = \frac{n}{2L}\sqrt{F/d}$$

where L is the length of the string and $n = 1, 2, 3,$ etc.

From the above equation it follows that:

- The fundamental mode ($n = 1$) has the lowest frequency.
- Longer strings produce lower frequencies.
- The frequency increases with tension.
- The frequency decreases if the linear density is higher, which means that thicker strings (of the same material) produce lower frequencies.

6.16 Problems and questions

1. (a) What is the longest wavelength a standing wave can have in a 0.6 m long open pipe?
   (b) What is the longest wavelength a standing wave can have in a 0.6 m long string clamped at both ends?
   (c) Is the frequency of the sound produced by the pipe and the string the same? Explain your answer.

2. (a) An open pipe has fundamental frequency of 50 Hz. List the frequencies of the first 3 harmonic overtones.
   (b) A closed pipe has fundamental frequency of 50 Hz. List the frequencies of the first 3 harmonic overtones.
   (c) Which of the two pipes is longer?

3. A certain instrument produces the following set of frequencies: 100, 200, 300, 400 Hz and so on.
   (a) Which frequency is the fundamental?
   (b) Which frequencies are the overtones?
   (c) Are the overtones harmonic? Why, or why not?
   (d) Which frequency is the second harmonic?
   (e) Which frequency is the second overtone?

4. A certain instrument produces the following set of frequencies: 100, 202, 303, 404,…Hz.
   (a) Are the overtones harmonic? Why or why not?
   (b) Is this a pure tone? Explain.
   (c) Which frequencies are the partials of this tone?

5. (a) Find the frequencies of the fundamental and the first four harmonics for a 2.4 m long open pipe.
    (b) How long should an open pipe be, to produce a fundamental equal to the second harmonic of pipe in part (a) above?
    (c) What would be the length of a pipe closed at one end, that would have the same fundamental as part (a) above?
    (d) What would be the first three harmonics of the pipe in part (c) above?

6. In terms of intervals (see table 5.2)
    (a) What does the second harmonic represent?
    (b) What does the third harmonic represent?
    (c) What is the interval of the fifth harmonic?

7. The ear canal is about a cylindrical tube about 0.03 m long, and ends at the ear drum.
    (a) Find the frequency of the fundamental mode of the ear canal.
    (b) Explain why our threshold of hearing (section 4.3) is lowest for frequencies between 2 and 4 kHz.

8. Here we will compare how standing waves and resonance frequencies would be affected in a hypothetical situation where our atmosphere was made of helium rather than air. The speed of sound in helium is 1000 meters per second.
    (a) If we pluck a guitar string, would you expect the wavelength of the modes to change?
    (b) Would you expect the frequencies of the modes to change?
    (c) How would the frequency of the sound reaching our ears compare to the vibrational frequency of the string?
    (d) What would be the wavelength of the fundamental mode in an open pipe of length equal to 1 m?
    (e) What would be the fundamental frequency of the open pipe in part (d)?
    (f) How would the frequency of the sound reaching our ears compare to the vibrational frequency of the standing wave in the pipe?

IOP Publishing

The Physics of Sound Waves (Second Edition)
Music, instruments, and sound equipment
Panos Photinos

Chapter 7

Sound equipment and components

7.1 Introduction

Sound systems have evolved dramatically in the second half of the 20th Century. The advances in electronic technologies and materials are probably more than people could have imagined in the 1950s. The main purpose of sound systems still remains the same: to output sound of good quality, at a power level suitable for the listening environment. The earlier sound systems used vacuum tubes, which are bulky, expensive to fabricate, produce a lot of heat, and have a limited lifetime, like ordinary incandescent light bulbs. Today vacuum tubes are still in use for specialized applications, and in guitar amplifiers by 'purists.' Otherwise, vacuum tubes have largely been replaced by the so-called **semiconductor** or **solid-state** devices, which is what we have in computers, cell-phones, digital cameras, and of course sound systems.

One of the major advantages of solid-state technology is the phenomenal reduction in size. A portable AM/FM radio nowadays is smaller in volume than a single vacuum tube. A key component in modern electronics is the **transistor**, which is a solid-state device that can use a *small* signal to control a large electric current provided by a large power supply. In other words, a transistor creates a larger replica or *amplifies the small signal*. In this chapter, we discuss the basic functions of sound equipment and components. To better understand the performance characteristics of these systems, we will introduce some concepts of electricity that will be useful in the discussion that will follow.

7.2 Concepts of electricity

7.2.1 Electric current, voltage and load

To introduce some of the basic concepts, we use the simple case of a voltage **source** and a **load**, as shown in figure 7.1. The voltage source, which could be a battery, an amplifier, a microphone and so on, drives an electric current *through* the load. The term load is generic, and can mean different things, depending on the nature of the source. For example, if the source is a wall outlet, the load can be a light bulb. If the source is an amplifier, the load can be a loudspeaker, if the source is a microphone, the load could be an amplifier, and so on.

By convention, the direction of the electric current is from the positive terminal of the voltage source, through the load, and back to the negative terminal of the source, as indicated in figure 7.1. If the voltage source provides a constant voltage, for instance if the source is a battery, then we speak of a direct current (DC) source. In this case, one terminal is labeled '+' and the other '−', or one is labeled '0' and the other '+' or '−'. In sound applications the magnitude of the voltage can be variable, and the polarity of the voltage can change as well, which means that the positive terminal becomes negative, then back to positive, and so on, in regular or irregular cycles. If the polarity of the voltage changes, then the direction of the current will change also, and we speak of a variable or alternating current (AC). The wall outlets in houses are AC sources. In AC sources one terminal is the 'neutral' which may or may not be connected to the ground, and the other terminal is the 'phase' or 'hot.' In power outlets the neutral should be connected to ground, and that is conventionally

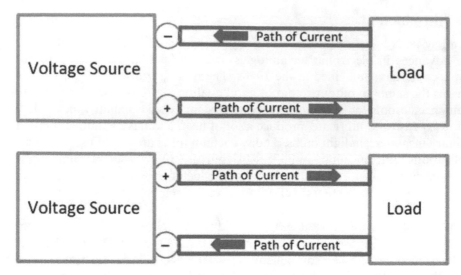

Figure 7.1. A load connected to a voltage source. The convention is that the electric current runs from the positive end of the source, through the load, to the negative end of the battery.

the 0 voltage. In North America the voltage between the phase and the neutral varies from about +170 to −170 V, 60 times per second, meaning that we have an AC of 60 Hz[1]. The current is usually carried by electrons, and the **electric current** is a measure of the rate of electron flow. The **voltage** is what provides the energy required for the electrons to move. As the electrons flow through the load, they slow down because they keep colliding with the atoms that make the wires or the other components the electrons may encounter along their path. The energy lost by the electrons during these collisions is converted to heat. The **resistance** between any two points (A and B) in any circuit is a measure of the hindrance encountered by the electrons as they move from point A to point B of the circuit.

7.2.2 Resistance, impedance, power and Ohm's law

Resistors are electronic components that have more or less constant resistance for both AC and DC voltages. A simple relation known as **Ohm's law** relates the voltage *across* a resistor to the current *through* the resistor:

$$\text{voltage} = \text{resistance} \times \text{electric current.}$$

Ohm's law tells us that for a given supply voltage, say 9 V, the product resistance × current must equal 9. For instance, we can have 1×9 or 2×4.5, or 3×3, and so on. Note here that as the resistance increases the current decreases and vice versa. Running more current means more loss of energy in collisions, that is, more energy lost as heat. Besides being a loss, heat in electronic devices can affect the

---

[1] The amplitude is 170 V. Commonly, the voltage is referred to as 120 V, AC. The 120 is the so-called root mean square (rms) value. The rms value is the amplitude (170) divided by the square root of 2. See section 2.9.2.

performance or even damage the unit. The current drawn/delivered to any unit must never exceed the values specified by the manufacturer. Resistance as described above is often called **Ohmic resistance**, to distinguish it from other mechanisms of hinderance to the electric current, which will be described later in this section. The term **impedance** is often used to include all processes that may hinder the flow of electric current.

The **power** delivered to a load is equal to the product of the voltage across the load, times the electric current through the load:

$$power = voltage \times current.$$

If the load is a resistor this power is 'dissipated', meaning it is converted to heat. This is essentially how a space heater works. A different load, such as a loudspeaker, would convert some of the power delivered into sound and some into heat.

The units used to measure the quantities introduced here are:
- Voltage is measured in **Volts** (V) or mV (1/1000 of 1 V).
- Electrical current is measured in **Amperes** (A) or mA (1/1000 of 1 A).
- Resistance and Impedance is measured in **Ohms** (Ω) or kΩ (1000 Ω).
- Power is measured in **Watts** (W) or mW (1/1000 of 1 W) or kW (1000 W).

7.2.3 Series and parallel connections

It is often the case to have more than one load connected to a source, for example two or three loudspeakers connected to one channel of an amplifier (see section 7.7). In the simplest case we can have two loads connected to a source. Figure 7.2 shows the two possible configurations. Here we will use some numerical values to make the discussion more concrete.

In figure 7.2(a) the two loads are connected to the voltage source **in parallel**. Here the 6 Volts of the voltage source are applied directly across each load. Let us apply Ohm's law for each load:

Load A: Resistance = 2 Ω; Current = 3 A; Resistance $\times$ Current = 2 $\times$ 3 = 6 V
Load B: Resistance = 1 Ω; Current = 6 A; Resistance $\times$ Current = 1 $\times$ 6 = 6 V

So, as one would expect, the voltage across each load is the same as the source voltage. The current through each load is different. Note that the current through

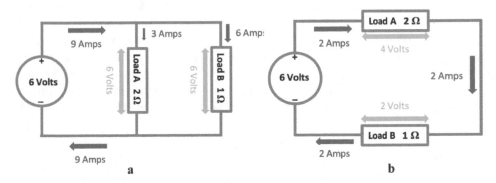

Figure 7.2. (a) Loads connected to the source in parallel. (b) Loads connected to the source in series.

the 1 Ω load is higher (= 6 A) than the current through the 2 Ω resistor (= 3 A). Note also that the sum of the currents through the loads (3 + 6 = 9 A) equals the current from the voltage source (= 9 A). In other words, the current supplied by the source is split between the two loads.

In terms of power, we have:

Load A: Voltage = 6 V; Current = 3 A; Voltage × Current = 6 × 3 = 18 W
Load B: Voltage = 6 V; Current = 6 A; Voltage × Current = 6 × 6 = 36 W

The power supplied by the voltage source is:

Source: Voltage × Current = 6 × 9 = 54 W which is equal to the sum of the power delivered to each load. It is important to note that when connected to the same voltage, the smaller impedance (load B in this example) draws more power. So, a *large impedance is a small load.*

In figure 7.2(b) the two loads are connected to the voltage source **in series**. Here the current through each load and the current through the voltage source are all the same. We also note that the voltages across each load sum up to the voltage of the source. In terms of power, we have:

Load A: Voltage = 4 V; Current = 2 A; Voltage × Current = 4 × 2 = 8 W
Load B: Voltage = 2 V; Current = 2 A; Voltage × Current = 2 × 2 = 4 W
Source: Voltage × Current = 6 × 2 = 12 W

Again, the sum total of the power delivered to the loads equals the power supplied by the source. Comparing the parallel and series connections, we note that *more power is drawn* by the two loads when they are connected *in parallel.*

7.2.4 Source and load impedance

In the above discussion we assumed that the voltage source is *ideal*, meaning that the source does not have any impedance. This is a useful model, but in reality, all components have some impedance, even plain wires. The following simple example will help illustrate the effect of this so-called *internal impedance* of the source. We can model this internal impedance as an extra resistor connected in series with the source. To simplify the discussion, we assume that the internal impedance is 10 Ω, and look at two cases: (a) Load = 2 Ω, and (b) Load = 100 Ω, as shown in figure 7.3. We see that in both cases, the voltage across the load is smaller than what the source can deliver (6 V). In case (a) the load gets only 1 V, that is only 17% of the source's voltage. The situation is much better in case (b) where we have 5.5 V, that is, 92% of the source's voltage[2]. Note that in both cases the sum of the voltages across the internal resistance and the load equals the voltage of the source. *As a rule, for audio systems the impedance of the load should be at the very least* 10 *times larger than the impedance of the source.* This is an important rule to keep in mind when connecting microphones, earbuds, loudspeakers and the like, as will be discussed later in this chapter.

---

[2] In terms of dB, we have 20 log 0.17 = −15 dB for case (a) and 20 log 0.92 = −0.7 dB. See sections 4.14.2 and 4.14.4.

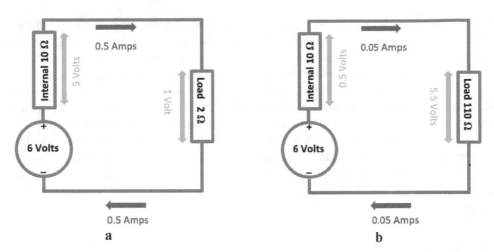

Figure 7.3. (a) Source impedance higher than load impedance. (b) Source impedance lower than load impedance.

7.2.5 Impedance of wires

In circuit analysis, it is commonly assumed that the wires have no effect on the circuit, which is not always the case. There are two points to keep in mind. First, one has to make sure that the wires are thick enough to handle the electric current. Otherwise, the wires will overheat or literally burn out. Second, one has to make sure that the impedance of the wires is much smaller (at least 10 times) than the impedance of the source and of the load. Otherwise, the impedance of the wires must be taken into account, and the result will be less voltage to the load. As the impedance of the wires depends on the length, the longer the wire, the higher the impedance, which imposes some *limits on the length* a wire can have. This point is particularly relevant when it comes to connecting microphones or loudspeakers, as will be discussed later on in this chapter.

7.2.6 Capacitance

As mentioned earlier, besides the Ohmic resistance, which affects both AC and DC signals, for AC currents there are two additional mechanisms that affect the flow of electric charge through a device. In the first mechanism the electric charge is temporarily stored. A device that stores electric charge is called a **capacitor**, which typically consists of two metallic films separated by an insulating layer. If a battery is connected across a capacitor, electric charge will flow from the battery to the films. The film connected to the positive terminal will end up with a net positive charge, and the other film with an equal amount of negative charge. Once the voltage between the films becomes equal to the voltage of the battery, the flow of charge stops, and so does the electric current. In other words, the capacitor acts as an 'infinite' resistance for DC. If an AC source is connected across a capacitor, the electric charges will flow back and forth from one film, through the connecting wires and through the source, and reach the other film. In other words, there will be an AC current through the wires and through the source[3]. This means that the capacitor conducts AC currents. The amount of charge that can be stored by a capacitor is *larger, the smaller the separation* or **gap** between the two films. The amount of charge stored by a capacitor connected to a source of 1 V is called the **capacitance**. A very interesting situation occurs when we change the size of the gap between the films[4]. Suppose that we have a capacitor connected to a 1 V battery. If we decrease the gap, the capacitance will increase, which means that the capacitor can now hold more charge. The result is that charges will move *from the battery to the films*, and we have an electric current. If we increase the gap, then the capacitance decreases, meaning that there is more charge on the films than they can hold. The result is that the excess charge will move *from the films to the battery*, so we will have an electric current in the opposite direction.

7.2.7 Inductance

A second mechanism that affects the flow of electric currents in circuits has to do with magnetism. Any wire carrying an electric current creates a magnetic field. To better understand how some sound equipment works, we need to discuss three situations where a wire interacts with magnetic field:

a. If a coiled wire carries electric current and happens to be in a magnetic field, then the magnetic field interacts with the magnetic field created by the electric current in the coil, and the result is a force on the coil. The force can be attractive or repulsive depending on the direction of the electric current relative to the direction of the magnetic field. As we will see in section 7.7, this force is what makes the cones of loudspeakers move, and produce sound.

b. A magnetic field moves in and out of a coil. In this case a voltage is *induced* at the two ends of the wire coil. This is the phenomenon of **induction**, and is exploited in making the so-called dynamic microphones, discussed in section 7.4.1.

c. Finally, if a wire carries an electric current, then the electric current 'feels' the magnetic field it produces around itself. The interaction between the current in the wire and the magnetic field the current builds around the wire is in a way similar to inertia. Inertia tends to *oppose any change* in the state of motion. In the same way, the magnetic field around a wire tends to oppose any *change* in the electric current that creates it. The result is that if somehow we try to increase (or decrease) the electric current in the wire, then the magnetic field around the wire will induce an electric current in the wire and the direction of the induced current will be such as to *oppose*, or *impede*, the *change* in the current that we are trying to impose on the wire. The induced change in the current will appear as a voltage at the ends of the wire. This phenomenon is called **self-inductance**, and is *higher the faster the change imposed* on the current, for instance in high frequency AC currents.

---

[3] The current through the separation between the films, that is the current through the gap, is negligibly small.
[4] See condenser microphones, section 7.4.1.

This property of **inductors**, as they are called, makes wire coils very useful in filtering out high frequencies, as discussed in the next section.

All three forms of impedance (resistance, capacitance and inductance) can be present together in any device (such as microphones, amplifiers, loudspeakers, *and* connecting cables). When connecting devices together, it is important to check the impedance of each device. As a general rule, the impedance of the voltage source must be at least 10 times smaller than the impedance of the load. Otherwise, the voltage across the load will be smaller than the voltage that the source can deliver, as discussed in connection with figure 7.3.

7.3 Filters

Filters are devices that can block a range of frequencies. The voltage signal to be processed is applied on the input side of the filter, and the filtered signal appears on the output side of the filter. The simplest filters use capacitors (to block low frequencies) and inductors (to block the high frequencies). Depending on the range of frequencies that are blocked, we have four basic types of filters: low-pass, high-pass, band-pass, and notch filters. An ideal **low-pass** filter will completely block all frequencies higher than a **cut-off** frequency. Similarly, an ideal **high-pass** filter will completely block all frequencies lower than a cut-off frequency. The **band-pass** filter will block all frequencies that are not within a certain range of frequencies. Finally, a **notch** filter will block only the frequencies that are within a certain range.

Figure 7.4 shows the response curves for 'ideal' filters. The vertical axis usually indicates **gain**, which is defined as the ratio of the output voltage to the input voltage,

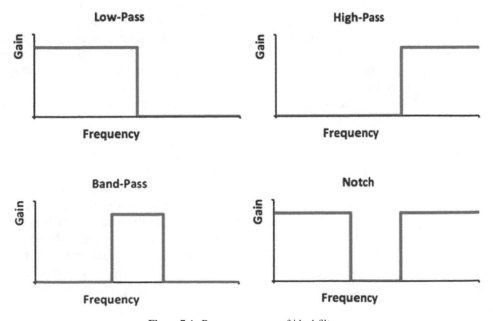

Figure 7.4. Response curves of ideal filters.

and is usually measured in dB (see sections 4.4 and 4.14.4). In figure 7.4, low gain means that the signal is blocked and vice versa. In practice, the response curves of real filters do not have corners, and the transitions from the passed to the blocked frequencies are wide and smooth[5].

The cut-off frequencies and the location and bandwidth of the pass-band and notch filters can be selected by the designer, and can even be 'tunable' as well, so one can have a knob that selects the cut-off frequencies, as is done with radios. In connection with loudspeakers, a very useful device is the **crossover** network of filters. As we will see in section 7.7, different loudspeakers respond to different frequency ranges. For example, a woofer responds only to the low frequencies, and the so-called tweeter responds only to high frequencies. The problem is that the output of an amplifier has all the frequency ranges together, and sending the high frequencies to the woofers or the low frequencies to the tweeters is a waste of output signal. The crossover network is a combination of filters that divide the output signal from the amplifier into two or more ranges (for example, low, mid-range, and high) and directs each range to the proper speaker: low frequencies to the woofers, high frequencies to the tweeters, and so on.

7.4 Microphones

7.4.1 Types of microphones

Microphones are devices that convert the pressure variation of sound waves into variations of voltage. A device that converts one physical quantity (such as pressure) to another (such as voltage) is called a **transducer**. The element or **capsule** of a microphone most commonly consists of a flexible **diaphragm** that is set in motion by the acoustic pressure of the sound wave. The next step is to convert this motion into electric voltage. One way is to attach a small wire coil to the diaphragm, and have the coil move in an out of a magnetic field. Recall from the discussion of the inductor (section 7.2.7b) that magnetic fields interact with moving coils, and this interaction *induces* a voltage between the two ends of the coil. If the coil is moving, a voltage replicating the motion of the diaphragm will develop between the two ends of the coil. This is the microphone signal. This type of microphone is called a **dynamic** microphone (see also section 7.8.5).

Another design uses a **piezoelectric** material. When pressure is applied between two sides of a piezoelectric material, a voltage develops between the two sides. In **piezoelectric** or **crystal** microphones, the diaphragm is attached to one side of a piezoelectric crystal, and the pressure applied by the diaphragm on the crystal creates a voltage that replicates the acoustic pressure from the sound wave. In another design, a metallic diaphragm is arranged parallel to a rigid metallic plate, the 'back-plate'. The pair of metal plates essentially forms a capacitor. A positive DC voltage (a battery, for instance) is applied between the back-plate and the diaphragm. As explained in section 7.2.6, if the diaphragm moves in response to an incoming sound wave, the gap between the plates of the capacitor (and its ability to

---

[5] See section 7.8.10.

store charge) will change accordingly. In other words, the charge stored by the capacitor will change, causing electrical charge to move (that is, causing an electric current to flow) to or from the battery. The current generated replicates the motion of the diaphragm. This type of microphone is the **condenser** microphone[6]. Condenser microphones need a voltage source, which could be a battery or an *electret*. Electrets are materials such as quartz or Teflon. These materials have a molecular structure that can separate positive from negative charges, so they act as a voltage source in the condenser microphone. Most electret condensers use Teflon. There are other types of microphones, including the ribbon microphone, and the carbon microphone, which will not be discussed here.

7.4.2 Directionality of microphones

One important characteristic of a microphone is the directionality or **directivity**. For example, suppose that we have a microphone and two loudspeakers, A and B. Speaker A is located on the axis of the microphone (at 0°) and B is at an angle from the axis (figure 7.5). The sounds from the two speakers are identical. One can then ask: at what distance should we place speaker B so that the signal picked up by the microphone is the same as the signal picked up from speaker A? Depending on the answer, we can have several types of microphones.

An **omnidirectional** microphone will pick up equally well from all directions, as long as the distance of B is the same as the distance of A from the microphone. In other words, we can place speaker B anywhere on a circle centered on the

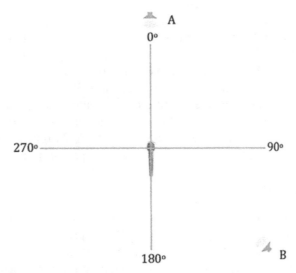

Figure 7.5. Two loudspeakers (A and B) producing identical sounds. The line 0°–180° is the axis of the microphone, which is located at the center of the figure.

---

[6] *Condenser* is an older term used for *capacitor*.

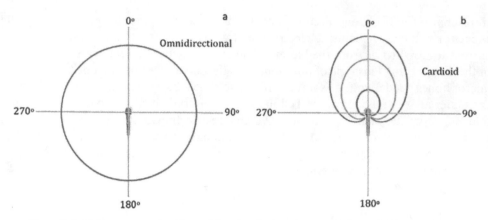

Figure 7.6. Pickup patterns for (a) omnidirectional microphones and (b) cardioid microphones.

microphone and get the same signal from the microphone. The so-called **pickup pattern** for an omnidirectional microphone is shown in figure 7.6(a). A very common pickup pattern is shown in figure 7.6(b). This is the **cardioid** pickup pattern. The microphone will produce the same signal from a speaker located anywhere on the red curve. The same applies for all locations on the green curve and on the blue curve, which is the farthest from the microphone. Note that in the forward direction (at or near 0°) the pickup is much better. If we set the speaker to the sides (at 90° or 270°) we need to bring speaker B much closer to the microphone to produce the same level of signal from the microphone. The microphone will hardly pick up any signal from a speaker set behind the microphone (at 180°).

Clearly the cardioid pattern favors the forward direction. This is an example of a **unidirectional** pickup pattern. An extremely unidirectional pattern is the so-called **shotgun** pickup pattern, where the microphone picks up only sounds coming along its axis (that is, at 0°). The pickup pattern of a microphone is basically determined by the design of the housing of the microphone capsule. Which pattern is best depends on the specific application. For example, on a theater stage, the omnidirectional would pick up the voices of all actors. For singing it may be best to use a cardioid, so as not to pick up sound from all nearby instruments. A shotgun pattern would work best for a reporter trying to record the voice of one person in a crowd, and so on.

7.4.3 Sensitivity of microphones

The sensitivity of a microphone tells us how much voltage is created by a sound of 1000 Hz that has rms acoustic pressure[7] of 1 Pa. Typically, the sensitivity is in the order of 2 mV/Pa for dynamic microphones and 20 mV/Pa for condenser microphones, which are more sensitive than the dynamic microphones. The sensitivity values are also given in dBv[8] using as reference an ideal microphone sensitivity of 1000 mV/Pa. For example, in terms of dBv, a sensitivity of 15 mV/Pa is 20 log (15/

---

[7] See section 2.9.2.
[8] As discussed in section 4.14.4, dBv means that we use 1 V as reference.

1000) = −36 dBv. From example 2 in section 4.14.1 we know that 1 Pa of acoustic pressure corresponds to 94 dB. So, in our example, the specification may appear as −36 dBv *re* 94 dB SLP, meaning referenced to sound pressure level of 94 dB[9].

The sensitivity characterizes the microphone response at 1000 Hz. The frequency response tells us about the response over the audible range. A commonly used brief specification is 'frequency response 20 to 20 000 Hz.' Unfortunately, this statement does not tell us if the response is the same over the stated frequency range. The proper specification regarding the response over the audible frequency range, which includes sensitivity, signal-to-noise ratio and harmonic distortion will be discussed in connection with amplifiers, in section 7.6.

7.4.4 Microphone impedance

Another important characteristic is the **microphone impedance**. If the impedance of the microphone is over 20 kΩ we have a high impedance microphone. Condenser and piezoelectric microphones are high impedance microphones. Low impedance microphones, such as a dynamic microphone, have impedance less than 1 kΩ, usually 50–500 Ω. When connecting a microphone to an amplifier, the microphone is the voltage source and the amplifier is the load. From the discussion of figure 7.3 we know that the impedance of the load must be higher than the impedance of the source. The input impedance of an amplifier is typically around 1–2 kΩ. So, when we use a cable to connect a high impedance microphone to an amplifier, the signal from microphone drops, and the drop affects mostly the high frequencies in the signal. This preferential drop of the higher frequencies comes from impedance due to the self-inductance of the cable. As discussed in section 7.2.7c, the inductive impedance is higher for higher frequencies. So, unless other precautions are taken, a high impedance microphone should not be connected to a cable longer than 6 m (about 20 ft). The low impedance microphones do not have this problem. Generally, the voltage output from a microphone is very weak, in the order of a few millivolts, and may require some amplification as a first step. The first stage is the so-called **pre-amplifier**, which brings the microphone voltage signal to the proper level.

7.5 The amplifier

The loudspeakers in a home sound system are typically in the range of 25–100 W, while the signals coming from a CD player, a tape player, and the like, are usually very low, in the order of mW. For example, the voltage output of a CD player is about 1 V and the current in the order of mA[10], meaning that the power output is about 1 V × 1 mA = 1 mW. This signal level needs to be amplified about a million times in order to drive a 100 W speaker. This amplification can be accomplished by increasing the current or the voltage or both. For safety reasons, high voltages are

---

[9] For digital microphones the situation is similar, except that the digital output voltage is referenced to the *full scale*, and indicated by dBFS. For example, if the output is 4-bit, then the full scale is 1111 meaning 15 in *decimal*. If the signal at 1 Pa is 0011, that is 3 in decimal, then the voltage ratio is 3/15 = 0.2. See section 8.1.
[10] See unit abbreviations in section 7.2.2.

not desirable, especially in home systems. Typically, the voltage to a loudspeaker is in the order of 10 V, although some concert speakers operate with voltages in the order of 100 V. Therefore, the power boost must come mainly from increasing the current by at least 1000 times. The so-called **power amplifier** (or **power amp** for short) does this final boost. Processing currents over a few mA generates heat, that is loss of power, meaning that any adjustments of the signal to be amplified must be done before the signal reaches the power amp; for example, adjusting the volume, the bass, or the treble. These controls occur in the pre-amplification stage, and can be relatively simple for most home systems, compared to studio sound recording and professional sound systems.

In most home sound systems, a **pre-amplifier** (**pre-amp** for short) is part of the **receiver**. The receiver in home stereos includes the **tuner** (for AM/FM radio) and a power amplifier, which feeds the loudspeakers[11]. In some home systems, and most stage and recording sound systems, the pre-amp is a separate unit. First the pre-amp brings the signal levels from all inputs to the right level for the power amplifier to handle.

In terms of frequency control, a simple pre-amp has knobs for tone control, that is, adjusting the mix of low and high frequencies, the bass and the treble, respectively. This is achieved by taking the signal through high-pass and low-pass filters before feeding the signal to the power amp. An **equalizer** uses a combination of filters to achieve finer control of the frequency bands. In this case, one can have three (or many more than three) frequency bands that can be adjusted separately.

Another function of the pre-amp is to add or mix signals. For example, adding the signals coming to the pre-amp from two or more different sources, and adjusting the proportions in the mix. For example, a DJ may have a microphone and two or more music sources (CD player, record player, smartphone, and the like). The DJ uses an **audio mixer**[12], which is a pre-amp that can mix the sound from all the inputs, and by adjusting the level of each, can make smooth transitions from one song to the next. Similarly, a pre-amp mixes the voice and the music in a karaoke machine. More elaborate units used in performances and recording studios can mix inputs from many (up to about a hundred in some cases) different inputs.

7.6 Amplifier characteristics

In this section we describe some characteristics of amplifiers. The concepts introduced here are also used in describing the performance of other components and devices, such as filters, microphones, loudspeakers and the like. A basic requirement for a good audio amplifier is to output a voltage signal that closely replicates the waveform of the input signal in terms of frequencies and amplitudes. For example, if the amplitude of a certain frequency in the input signal doubles in 1 s, then the output signal should do the same, otherwise the sound comes out distorted. To avoid **distortion** the amplifier must be **linear**[13]. Typically, the

---

[11] See discussion of AM/FM bands in section 1.6.
[12] See further discussion in section 7.8.7.
[13] See example in section 10.7.5.

components of amplifiers, mainly the transistors, have a certain input range for which the response is linear. If the input exceeds this range, the output will be distorted.

A response curve is a plot of the magnitude of the output signal versus the input signal; meaning that if we know the input value (horizontal axis) we can read the output from the vertical axis. Figure 7.7 shows an 'ideal' response (green line) and a more realistic response curve (red line). We note that for small input signals (less than about 0.04 V in this example) the red curve is fairly linear. For an input of 0.02 V, the output is 2 V whether we use the red or green line. This is not the case for higher input signals. If the input is 0.1 V, the red curve reads about 7.5 V and the green line reads 10 V.

A non-linear response adds harmonics (which are not present in the input signal) to the fundamental frequency of the tone. The fraction of the power of the added signal (from the harmonics) to the signal of the fundamental is the **total harmonic distortion (THD)**. The THD is measured in %, and a typical manufacturer specification may read: less than 1% THD. Of course, the higher the percentage, the lower the quality of the amplifier. A quality unit should have THD less than 0.1%.

The ratio of the output to the input voltage is the **voltage gain**. For example, in the linear range of figure 7.7, a 0.02 V input gives an output of 2 V. So, the gain in this

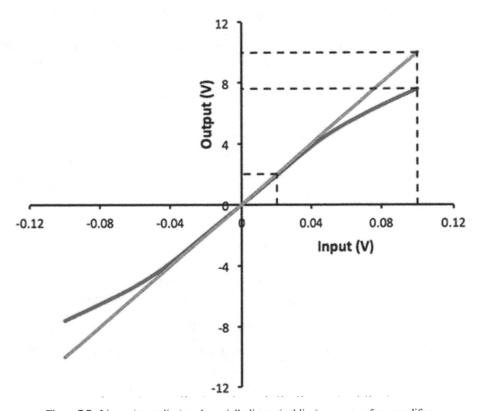

Figure 7.7. Linear (green line) and partially linear (red line) response of an amplifier.

case is $2/0.02 = 100$. Similarly, we can define the **power gain** as the ratio of the power output to the power input. It is common to express the gain in decibels (dB), as discussed in section 4.4. A power gain of 20 dB means that the amplifier boosts the power of the input signal by a factor of 100, and a gain of 60 dB means a power boost by a factor of 1 million. Of course, a high gain is desirable, but most importantly the gain should be the same for a wide range of frequencies, and preferably covering the entire audible range, from about 20 Hz to 20 kHz.

Figure 7.8 shows a frequency response curve. Note that the frequency axis is logarithmic, that is, the steps represent factors of 10, (see section 4.2). The first vertical grid line after 10 Hz corresponds to 20 Hz, the next to 30 Hz, and so on. The response curve in this example levels out at 20 dB, and *rolls-off* at the two ends of the frequency range. The green dotted line marks the 17 dB gain level, and the two green dots indicate the points where the gain drops to 17 dB, that is, the gain drops by 3 dB from the highest value. In terms of power, a 3 dB drop means that the power drops by a factor of 2, and this happens at 50 Hz in the low frequency side, and 15 000 Hz in the high frequency end of the diagram. These two frequencies define the useful range of the amplifier, the so-called **bandwidth** of the amplifier. Usually, the specification is given as '30 Hz–16 kHz ± 2 dB', meaning that in the range of frequencies stated, the gain for some frequencies may be at most 2 dB above or below the average. A 2 dB difference means a factor of 1.6 in power[14]. In other words, some frequencies are amplified 1.6 times more than the average (+2 dB) or 1.6 times less than the average (−2 dB).

All electronic devices are susceptible to **noise**, usually from wires, from the supply voltage, from each other, and other sources. This is how we can tell that a stereo is

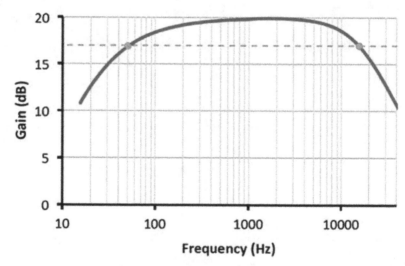

Figure 7.8. Frequency response curve for an amplifier. The green dots at 50 and 15 000 Hz indicate the frequencies where the gain drops by 3 dB (50%).

---

[14] See example 2 in section 4.14.2 for this calculation.

on, even if there is no music playing. Naturally for any output signal to be useful, it should be louder than this unavoidable noise. The **signal-to-noise ratio (SNR)** compares the maximum output signal to the noise signal. For example, SNR = 100 dB means that the maximum output is 10 billion times higher than the noise. Recall from section 4.4 that a sound power level of 100 dB above any audible noise brings us close to the upper limit of what our ears can handle without damage.

Related to the noise level is the **dynamic range** of the amplifier, defined as the difference (in dB) between the maximum power level and the noise. For example, if the maximum output is 100 dB and the noise is 30 dB, then the dynamic range is 100−30 = 70 dB. A high dynamic range allows a better separation of what is supposed to sound louder in a piece of music, and what is supposed to sound softer, yet remain audible above the noise level.

An amplifier converts the power supplied by a power source to a useful audio signal. Audio amplifiers are not very energy efficient devices. The **efficiency** is measured as the ratio of the output power of the amplifier to the power the amplifier draws from the electric outlet or battery. The efficiency typically ranges from 10% to 50%. For instance, an amplifier delivering 50 W at the output and 20% efficiency actually uses 250 W total power. It outputs 50 W to the loudspeakers and 200 W as heat. The output power rating of audio amplifiers measures the power of the audio signal at the output of the amplifier, not the loudspeakers. For home units it can range from about 20 W to over 100 W, and much higher for professional units used in concerts. This power is delivered to the speaker system, and a very important step is to match the power rating of the amplifier to the power required to run the speakers. Generally, the power rating of the amplifier should be equal to the power rating of the speakers. If the amplifier rating is lower, then the speakers will not perform to their best. On the other hand, if the amplifier rating is much higher than the loudspeaker's rating, the loudspeakers will be damaged.

Another amplifier specification is the **damping factor**, which is the ratio of the impedance of the load divided by the output impedance of the amplifier. The output impedance of the amplifier is low, in the order of 0.1 Ohms. So, if the impedance of the load is 4 Ohms[15], we have a damping factor of 4/0.1 = 40. If the impedance of the load is 8 Ohms, then the damping factor is 8/0.1 = 80. This tells us that the damping factor in the specification must state the impedance of the load. Generally, a large damping factor is preferable[16].

7.7 Loudspeakers

The loudspeaker's function is in a way the reverse of the microphone's function. The loudspeaker converts an electrical signal from the amplifier to a sound wave. The microphone converts a sound wave into an electrical signal. Because of this

---

[15] The impedance of the load includes an inductive component from the sound coil discussed in the next section. So, the impedance of the load depends on the frequency of the signal.
[16] See also section 7.8.4.

similarity, one can plug-in headphones or ear buds to the input of an amplifier and use them as microphones. In some cases, as in two-way radio units, a speaker is used both as a microphone and as a loudspeaker. The most common type for home audio systems is the **dynamic loudspeaker**, which will be described in this section. The output signal from the amplifier is connected to the voice coil, which consists of several turns of thin wire, usually copper covered with thin insulating coating. The voice coil is suspended in a way that allows it to move back and forth in the *bore* of a magnet. The input signal from the amplifier produces an electric current through the coil, which in turn produces a magnetic field around the coil. The magnetic field of the coil interacts with the magnet (section 7.2.7a) and this interaction causes the coil to move back and forth in a pattern that mimics the input signal. The sound coil is rigidly attached to a diaphragm that sets the surrounding air into vibration, and produces a sound wave that mimics the input signal. The diaphragm is usually cone shaped, and is attached to the body of the speaker by a flexible suspension, as shown in figure 7.9.

The range of back-and-forth travel of the sound coil defines the **excursion** of the speaker. At low frequencies, the diaphragm moves slowly, so the rate at which energy is delivered to the surrounding air is slow. Recall that power is the rate at which energy is delivered (see section 1.5). This means that to achieve a certain level of power at low frequencies, a larger amount of air must be moved by the diaphragm

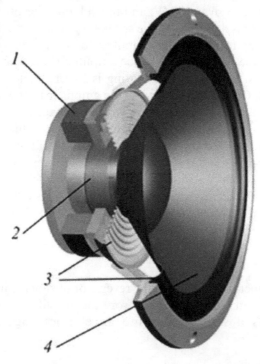

Figure 7.9. Cutaway view of a loudspeaker. 1. Magnet. 2. Sound coil. 3. Suspension. 4. Diaphragm. Image credit: Svjo (Own work) [CC BY-SA 3.0 (http://creativecommons.org/licenses/by-sa/3.0)].

compared to higher frequencies. Low frequencies need a large diaphragm, with long excursion. On the other hand, larger diaphragms cannot move fast enough to follow the high frequencies effectively. This is why a loudspeaker system usually includes three different speakers, or **drivers** as they are often called. A small driver for the high frequencies (over 6 kHz) is called the **tweeter**. The **mid-range driver** covers mid-range frequencies (about 1–6 kHz). The **woofer** covers the low frequency end. All three drivers are usually housed in the same enclosure, and cover the entire range of audible frequencies in most cases. A **sub-woofer** is a low frequency (below 200 Hz) driver and has its own box and own separate amplifier, and is usually placed on the floor facing down. The size of the cone is smallest for tweeters (2–5 cm diameter) and about 20–50 cm for woofers and subwoofers. As the different drivers work with different frequency ranges, a **crossover** network of filters is used to route the different frequency ranges to the appropriate driver[17]. In some applications, covering the entire audible frequency range is not so important (for example, computers and TV sets). These units can use a single mid-range speaker (10–20 cm cone) and usually have a double cone to enhance the higher frequencies.

Amplifiers have labels specifying the impedance of the speakers to be used, usually 4–16 Ohms. Recall from section 7.2.3 that a load (which is the loudspeaker in our case) of low impedance draws more current from the voltage source (which is the amplifier); meaning that we should not connect a loudspeaker with impedance lower than the minimum specified for the particular amplifier, other-wise the speaker will be driven beyond its excursion range, which will cause distortion of the sound. Worse yet, drawing high current from the amplifier causes heating of the transistors, which in excess can damage the amplifier. This can happen when two or more speakers are connected in parallel[18] to the same output terminal of the amplifier.

Loudspeakers are not very efficient devices. Typically, they convert to sound only about 1%–5% of the power delivered by the amplifier. In other words, if 100 W of power are delivered to the loudspeaker, the sound wave coming from the loud-speaker will have power of 1–5 W, and 95–99 W are converted to heat. This means that loudspeakers tend to get hot! Transferring the energy from the cone to the surrounding air is inherently inefficient, and that is the main reason for the low efficiency of loudspeakers.

Headphones are usually miniature dynamic type speakers. They require very low power, in the order of mW, so they can be driven by a pre-amp, and no power amplifier is needed. As they are at close proximity to the eardrum, they do not require a large diaphragm for low frequencies. So, their output frequency range can go down to 40 Hz or less. The most important concern with earphones is the level of sound reaching the ear. As they get so close to the eardrum, the entire output power is delivered into the ear canal, and the sound level can easily reach 80 dB, or even exceed 100 dB, which can cause ear damage, as discussed in section 4.4.

---

[17] See section 7.3.
[18] See figure 7.3.

7.8 Further discussion

7.8.1 Amplifier power specifications

Amplifier specifications can be misleading at times. One must keep in mind that for AC signals, the power is also oscillating from a minimum value to a peak value. So, it is the average value of the power that is important. The peak value is actually larger than the average value, so, listing the peak value (or not stating whether it is peak or average) may be misleading the consumer. If the amplifier has more than one channel then it is important to know whether the power listed is total power or power per channel. In the case of two channels, the listing may be abbreviated as '100 Watts × 2', meaning 100 W per channel. Note that the power delivered depends on the load. A load of low impedance draws more power. For example, the power output to a 4 Ohm speaker may be 120 W but for an 8 Ohm speaker, the output power may be only 60 W. This means that the power specified must refer to the speaker impedance, for example '50 W to 4 Ohms'. The amplifier must be able to output the specified power level continuously, and the appropriate specification should clearly indicate 'driven continuously'. Also, the power must refer to the entire frequency range, not just a single frequency, say 1000 Hz. Finally, the THD should refer to the power level stated. Not listing the THD is a sign of low quality.

7.8.2 Reducing noise pickup in cables

The wall outlets carry an AC voltage (60 Hz in North America) and this frequency, and its multiples (120, 180 Hz and so on) affect many audio components. This is the hum we hear from amplifiers when nothing is playing. Besides the 60 Hz noise, the line may carry other noise, but in an average amplifier, the sound signal is much louder than the noise (SNR ratio in section 7.6). What is more common, is the noise that is picked up by microphones, guitar cables, and the like. The usual approach is to use 'shielded' wires. In this case, the signal carrying wire or wires are surrounded by a conducting sleeve. For low frequency applications the shield is a braided wire mesh. For higher frequencies, the shield is a conducting foil. The 'balanced pair' connection is a very simple and effective technique to reduce noise, and uses two wires for the signal with the neutral (0 Volts) wire wrapped around the pair. The key point is that the signals in the two wires are mirror images of each other, as shown in figure 7.10. The noise that is picked up by the cable (red spikes in the figure) has the same sign (positive or negative) in both lines. Once the wires get to the destination device, one could take the voltage *difference* between the two lines. Since the noise is

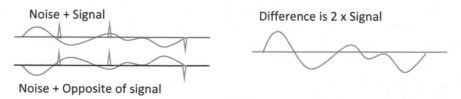

Figure 7.10. Balanced pair wiring. Red spikes indicate noise picked up by wires, and affect both wires in the same way, that is, they have the same sign, positive or negative.

the same in both wires, it *cancels out*. The op-amps discussed in section 7.8.6 can make the negative replica, and can also take the voltage difference between the two wires. This is done by connecting one wire to the inverting input pin, and the other wire to the non-inverting input pin of the op-amp.

7.8.3 Loudspeaker enclosures

Loudspeakers used in home systems usually have the three drivers (tweeter, mid-range, and woofer) housed in the same enclosure. The enclosure must be very rigid, with braces on the inner side of the enclosure that keep the sides from vibrating. The enclosure is usually filled with soft material to absorb the sound wave produced by the backside of the cone, which is 180° out of phase compared to the wave from the front side (see section 2.8.3). In some designs, the loudspeaker enclosure is almost airtight; this is so-called air-suspension, or acoustic suspension. Another design includes an open tube in the enclosure. This tube is usually mounted at the bottom of the front side, and helps to extend the system's performance at very low frequencies. This so-called bass-reflex design can 'tune' the phase of the sound produced by the back-side of the diaphragm, so that the sound coming out of the tube is in-phase with the sound coming from the front of the loudspeaker. To understand how this works, consider the enclosure plus tube as an air cavity, described in section 6.14.1. The resonance frequency of the cavity can by adjusted by changing the length of the neck (the tube in this case). If the resonance frequency is chosen to be just below speaker/woofer output, then from figure 6.7(b) we will have a phase difference of 180°. What this means is that the sound from the tube will have a phase difference of 180° from the sound in the enclosure. Now the sound from the front of the speaker has also 180° phase difference from the sound in the enclosure, meaning that the sound from the tube is in phase with the sound from the front. The opening of the tube is flared, to smooth out the change in impedance, which helps minimize the reflection of sound back to the enclosure, and maximize the transmission in the forward direction (see section 2.4.3).

7.8.4 Connecting loudspeakers

Loudspeaker connecting points on stereo amplifiers and connecting cables are color-coded and for good reason. The signal from the amplifier to the loudspeakers is an alternating (AC) signal, meaning that the 'plus' and 'minus' alternate with time. If correctly connected, the diaphragm of both loudspeakers should move say in the forward direction, when the signal is positive, and move backwards when the signal becomes negative. If the wire connections of one speaker are reversed, then the diaphragm of one speaker will be moving forward while the diaphragm of the other speaker is moving backwards, so, the motion of the two diaphragms will be 180° out of phase. If we have a certain sound wave coming from both speakers simulta-neously the sound waves from the two loudspeakers will cancel each other to some extent, and the sound level will be reduced.

As the impedance of loudspeakers is typically in the order of a few Ohms, and the output impedance of the amplifiers is usually much lower (about 0.1 Ohms), meeting the × 10 rule discussed in section 7.2.4 is not a problem. What can be a problem is the added resistance from corroded wires and corrosion of the contacting surfaces at the connection points. Another problem comes from the resistance of the wires. The resistance of 12 gage (2 mm diameter) copper wire is 0.005 Ohms per meter. So, the resistance of a 40 m wire is 0.2 Ohms. This wire would be long enough to connect a loudspeaker 20 m away from the amp (the wire needs to be double). If the impedance of the speaker is 2 Ohms, then the impedance of the wire is 1/10th that of the loudspeaker, just at the limit of the × 10 rule. Here we assume that the impedance at the output of the amplifier is negligible. The output impedance of the amplifier is usually around 0.1 Ohm. Now, from the loudspeaker side, the current though the coil must be able to build up and discharge quickly, so that the motion of the coil can follow the signal from the amplifier *faithfully*. For this to happen, we need the impedance seen by the coil, which is the sum of the impedance of the amplifier plus the resistance of the wire connecting the loudspeaker to the amplifier, to be low. As we found above, the wire resistance can be 0.2 Ohms, which is twice

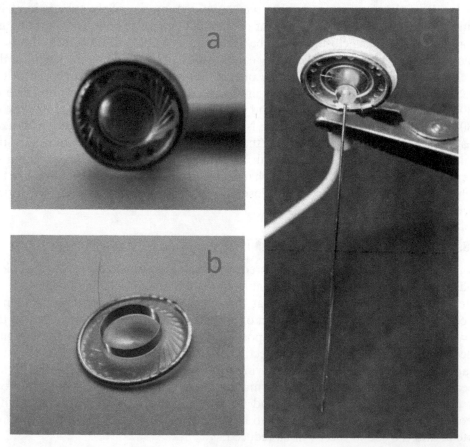

Figure 7.11. Parts of an earphone. (a) Front cover removed. (b) Membrane with coil. (c) Needle suspended by permanent magnet.

the impedance of the amplifier. The conclusion here is that we need to minimize the resistance of the connecting wires, and make sure that the wires and connecting terminals are clean.

7.8.5 Earphone and dynamic design

The dynamic design is used both in microphones and loudspeakers. Figure 7.11 shows the main parts of an earphone. The cone here is a transparent film in figure 7.11(a) and the copper wire coil is attached to the bottom of the cone (figure 7.11(b)). The coil fits in the circular groove surrounding the permanent magnet shown in figure 7.11(c). If a voltage signal is applied between the two ends of the wire, the magnetic field generated by the coil interacts with the permanent magnet, and the coil starts moving back and forth. This motion sets the cone into vibration. This is how the speaker works.

On the other hand, if the membrane is set to vibration by an acoustic wave, the coil moves back and forth in the magnetic field and this motion generates a voltage difference between the two ends of the coil wire. This is how a dynamic microphone works.

7.8.6 Operational amplifiers (op-amps) and the pre-amp

The integrated operational amplifier is essentially a circuit of 30 or more components, mostly transistors with a few resistors and capacitors, all 'printed' on a silicon chip. Although the number of components may seem large, the actual device is extremely simple. Typically, the op-amp has 8 or 14 pins, some of them are not connected at all. The standard pins are:

- two pins for signal input, indicated by '+' (the 'non-inverting') and '−' (the 'inverting')
- one pin for output (out) and
- two pins, $+V_s$ and $-V_s$, for the voltage that runs the amplifier, typically ±5 or ±12 V. Usually these pins are *not shown in schematics*.

All voltages are referenced to ground. The *absolute value* of the output signal *cannot* exceed the voltage supply $+V_s$. In practice, this means that if the output exceeds this value, the output signal will be flat, and the information is lost. This distortion is called **clipping**.

Figure 7.12(a) shows a 14-pin op-amp, and (b) the inverting amplifier circuit. The output voltage is determined by the ratio of the two resistors:

$$V_{out} = \frac{-V_1 R}{R_1}$$

This circuit is essentially the **pre-amp**. In this figure we used low resistances for simplicity. In practice, the input resistors are in the $k\Omega$ range, and the top, or feedback resistor, in the order of hundreds of $k\Omega$. If one of the resistors is variable, then we have a 'volume control' knob. Note that the output has the opposite sign of the input (hence the name *inverting*), which is not a problem because we can use a second inverter to restore the sign of the output signal if necessary.

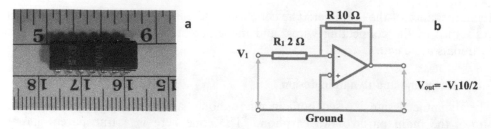

Figure 7.12. (a) A 14-pin op-amp. (b) Inverting amplifier circuit.

An important property of the op-amp is that it has very high input impedance (theoretically 'infinite', in reality 1 million Ω minimum) and very low output impedance (50 to 200 Ω). Very high input impedance means that we can connect at the input a high impedance signal source, like a condenser microphone, a turntable cartridge or a magnetic tape head, and easily meet the '10 times' rule mentioned in section 7.2. Besides voltage amplification and mixing, op-amps are used in analog-to-digital and digital-to-analog converters (see section 8.3), filters, and timers, to mention only a few.

7.8.7 Audio mixer

If we add a second signal to the input of the op-amp in figure 7.12(b), as shown in figure 7.13(a) then the voltage at the output becomes:

$$V_{out} = -\left(\frac{V_1 R}{R_1} + \frac{V_2 R}{R_2}\right)$$

With the resistor values used in figure 7.13(a) this means that $V_{out} = -(5V_1 + 2V_2)$.

We can add a third voltage to the input, the result is that the output will be a mix of the three voltages at the input. The mixing proportions are determined by the resistors at the input. These resistors could be variable; in such a case the user can mix the input signals at will. This is basically the **audio mixer**, shown in figure 7.13(b).

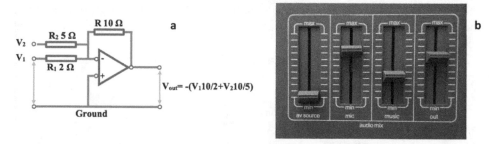

Figure 7.13. (a) Mixing two signals. (b) Audio mixer.

7.8.8 Electric guitar pickup

Special types of microphones, the so-called pickups, are used to amplify the sound of electric guitars. The pickup is similar in principle to the dynamic microphone. A coil

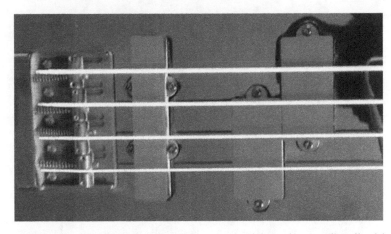

Figure 7.14. Pickups on an electric bass guitar. On the right we have a split-coils pickup.

is wrapped around a magnet placed just below the metallic string. As the string vibrates, the motion of the metal causes a change in the magnetic field seen by the coil. The changes in the magnetic field induce a voltage between the two ends of the coil. The change in the voltage has the same frequency as the vibration of the string. This weak signal is sent to the pre-amplifier, as in the case of some ordinary microphones. The coils of the pickup are susceptible to the 60 Hz noise from the wall outlets which appears as a 'hum.' To reduce the noise, it is common to use double coils of opposite magnetic polarity. The signal from the string has opposite polarity in the two coils. The 60 Hz noise has the same polarity in the two coils. If the wires at the output of the two coils are *cross-connected*, in other words, if we take the *difference* between the signals from the two coils, the useful signals will *add* and the noise from the two coils will *cancel*. This cross connection is called '**bucking**' in electronics, and this type of pickup is called the '**humbucker**'. Figure 7.14 shows one pickup near the bridge (left) and two split-coils on the right, on an electric bass guitar.

Guitar pickups are usually divided into active and passive. The difference is that the active has essentially a pre-amp, so the signal is at higher level. This can cause a problem if the gain of the pre-amp is too high. In such case the output signal from the amplifier will saturate (see 'clipping' in section 7.8.6) and the sound will be distorted. To prevent this from happening, the signal from the active pickup may have to be reduced or *attenuated*. Most amps will have two inputs. One marked 'passive' or '0 dB' and the other 'active' or '−10 dB'. The number in dB indicates the attenuation, and may range around −10 to −15 dB.

7.8.9 An example of filtering

Figure 7.15 shows how to use the response graphs of a filter to find the waveform of the filtered signal. We will use the waveform analyzed in section 3.10.3. The filtering actually affects the amplitude spectrum of the signal. So, the first step is to get the amplitude spectrum of the signal, which is shown in figure 3.18. The next thing to do is to multiply the response graph of the filter times the amplitude spectrum. Here we have two filters, a low pass and a high pass. The response shows that the low pass

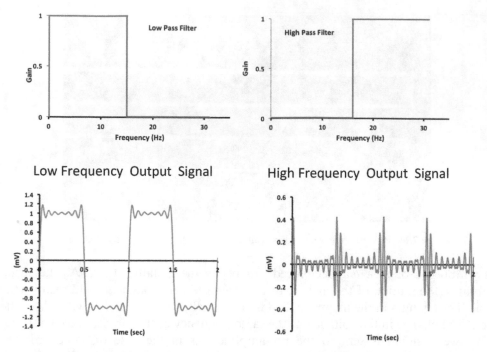

Figure 7.15. Filtering the square wave analyzed in section 3.10.3.

has cutoff at 15 Hz, in other words, the output of the filter is zero for any frequency above 15 Hz. The high pass has cutoff at 17 Hz, so all frequencies below 17 Hz will not appear at the output of the filter. In other words, the amplitude spectrum of the low pass filter is essentially the first eight peaks (up to 15 Hz) of the amplitude spectrum of the input waveform. In the same way, the amplitude spectrum of the high pass filter consists of the last eight peaks (17 Hz and up) of the amplitude spectrum of the input waveform.

To find the waveform output by each filter, we should add a term for each frequency component, as was done in section 3.10.1. The results are shown in figure 7.15 for the low pass and the high pass filters, respectively.

7.8.10 Non-ideal filters

In the filter response graphs shown in figure 7.4 the cut-offs are very steep with sharp corners, which is an idealized situation. In practice, the *roll-off* happens gradually, and over a wide range of frequencies, as shown in figure 7.16. For example, the roll-off for the low-pass filter starts around 100 Hz and extends to about 5000 Hz, in other words, a fair amount of the 1000 Hz frequency can still pass through the filter, which can be a problem. (Note that the frequency axis is logarithmic.) The roll-off can be shortened using combinations of many inductors and capacitors. The problem is that while these combinations can achieve steeper roll-off, they may also introduce a distortion of the signal. This *delay distortion* happens because different frequencies may take a longer time to get through the filter. Operational amplifiers can be used in filter designs, the so-called *active filters*, which generally have steeper roll-off. Another approach is to use

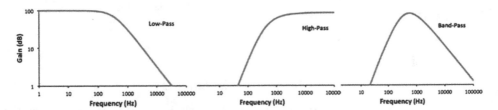

Figure 7.16. Response curves for simple non-ideal filters, typically involving one or two capacitors and inductors.

digital filters, meaning that the signal must be converted first into digital form, then processed by a computer. Processing usually involves various schemes of averaging the data. The signal is then converted from the digital form to a filtered waveform, as will be discussed in the next chapter. The problem here is that the two conversion steps, and the *number* of data points required may cause considerable delay, while the simple passive filters can do the filtering 'on-the-fly.'

7.8.11 Audio feedback and screech

When using an amplifier in a lecture hall or a performance, the goal is to pick up most of the sound from the speaker or performers, and minimize the pickup from other sources, especially the sound from the loudspeakers. Amplifying the sound from the loudspeakers can result in a loud **screeching** sound. As illustrated in figure 7.17, the sound signal is making a loop from the microphone to the amplifier to the loudspeaker and back to the microphone. So, we have what is called a **positive feedback loop**, meaning that the output signal is fed back to the input, so it keeps growing. The frequency of the signal that is amplified by the feedback loop is determined by a number of factors, which include any frequencies that are preferred by the microphone, the amplifier, the speaker, the room, and the distance between microphone and loudspeakers. Reducing the gain of the amplifier would eliminate the screech. On the other hand, reducing the gain may defeat the purpose of the amplifier. Using directional microphones pointing away from the loudspeakers is a good solution. Another solution is to use an equalizer to reduce the frequency produced by the feedback loop.

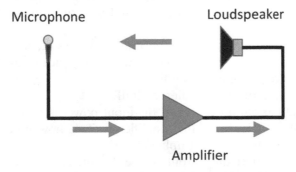

Figure 7.17. Path of audio signal in a positive feedback loop.

7.9 Equations

Ohm's law for AC voltages

For AC circuits Ohm's law reads

$$V_o \sin(2\pi ft) = Z I_o \sin(2\pi ft + \phi)$$

where Z is the impedance connected to the AC voltage source, V_o and I_o are the amplitudes of the voltage and current, respectively, f is the frequency, and ϕ is the phase difference between the current and voltage. We will ignore the phase difference here. For a loudspeaker the impedance is $Z = \sqrt{R^2 + (2\pi fL)^2}$, where R is the resistance of the sound coil and L the inductance of the sound coil, measured in units of Henry (H). Note that the impedance increases with increasing frequency, therefore the load gets 'lighter' at higher frequencies.

7.10 Problems and questions

1. Suppose we have a source of constant voltage and two loads of impedance 5 and 10 Ohms.
   (a) Which load draws more current from the source?
   (b) Which load draws more power?
   (c) Thick short wires allow current to flow freely without impedance. Suppose we connect a thick wire between the terminals of the voltage source. Will this load draw high or low current?

2. Ohm's law tells us that: voltage = resistance × electric current.
   By definition: power = voltage × current.
   We can combine the two equations to get
   power = (voltage)^2/ resistance or power = resistance × (current)^2.
   (a) Could the electric current be negative?
   (b) Could the voltage be negative?
   (c) Could a resistor have a negative Ohm value? Can you think of a mechanism that would make the resistance negative?
   (d) Can you think of a resistor that comes close to 0 Ohms? Would such resistor dissipate power? If yes, then where does this energy go?
   (e) Can the power dissipated by a resistor be negative?

3. (a) Which type of microphone would be more suitable for a singer in a band?
   (b) Which type of microphone would be more suitable for a theater stage?
   (c) Which type of microphone would be more suitable for a news broadcaster in a TV studio?

4. The power gain of an amplifier is 20 dB.
   (a) Find the output power, if the input power is 1 Watt.
   (b) If the efficiency of the amplifier is 50%, how much power must be supplied to run the amplifier?
   (c) Suppose that the output of the amplifier is connected to an 8-Ohm loudspeaker. The efficiency of the loudspeaker is 10%. How much

sound power does the loudspeaker put out? What happens to the rest of the power supplied by the amplifier?

(d) Suppose that we connect a 4 Ohm loudspeaker instead. Will this loudspeaker draw more or less power? Explain.

5. Suppose that we have a non-linear amplifier that has power gain of 100 dB. Explain how the non-linearity would affect the performance of the amplifier under the following conditions:

(a) When the volume setting is very low.

(b) When the volume setting is very high.

(c) Which characteristic of the amplifier (e.g., the gain, the total harmonic distortion, etc) would most likely be affected by the non-linearity?

(d) Which frequency ranges (bass, midrange, and treble) would be affected the most by the non-linearity?

6. Assume that an earphone outputs 0.002 Watt into one's ear, and that the ear's opening is 1 cm^2.

(a) How many square centimeters in one square meter?

(b) Use the result of part (a) to find the intensity of the sound (see section 1.5).

(c) How does the above intensity compare to the threshold of pain? (see section 4.3).

7. Suppose that the voltage output of an amplifier is a pure sinusoidal signal. The amplifier is connected to a loudspeaker.

(a) What is the average value of the signal over one cycle?

(b) Does the result of part (a) tell us anything about the power delivered to the loudspeaker?

(c) What would be a better measure of the power delivered to the loudspeaker? *Hint*: Look at problem 2 of this section.

(d) Suppose that the amplitude of the output signal is doubled. How would this affect the power delivered to the speaker?

8. Typically the power delivered by the amplifier to an ordinary loudspeaker is in the order of 10 Watt, and the signal voltage is around 10 V. The power delivered to earphones is in the order of 10 mW, and the signal voltage is around 1 V.

(a) How does the current through the loudspeaker compare to the current through the earphone?

(b) How does the impedance of the loudspeaker compare to the impedance of the earphone?

IOP Publishing

The Physics of Sound Waves (Second Edition)
Music, instruments, and sound equipment
Panos Photinos

Chapter 8

Analog and digital signals

8.1 Introduction

In everyday life we use the **decimal system** to represent numbers. The decimal system uses the **digits** 0–9 and is based on the powers of 10, that is, 1, 10, 100, 1000, and so on[1]. For example, in the decimal system the four-digit number 2193 means (reading from right to left) 3×1, plus 9×10, plus 1×100, plus 2×1000. There are many number systems, based on powers of numbers other than 10. As will be explained below, the so-called **binary system**, which is more suitable for computers, is based on powers of 2, that is, 1, 2, 4, 8, and so on. For example, in binary the number 1101

---

[1] Note that any number raised to the power 0 equals 1. For example, $10^0 = 1$ and $2^0 = 1$, and so on.

means (reading from right to left) 1×1, plus 0×2, plus 1×4, plus 1×8, which equals $1 + 0 + 4 + 8 = 13$ in decimal notation.

The binary digit or **bit** for short is the basic unit in the representation of binary numbers, and the system uses the digits 0 or 1 in each place. Using one bit we can represent the number 0 or the number 1. Using two bits we can represent four numbers (including 0), namely the number 0 (as 00), the number 1 (as 01), the number 2 (as 10), and the number 3 (as 11). Obviously, more bits are needed to represent numbers larger than 3. Following are the first eight (including 0) decimal integers in binary representation using three bits, with the decimal number in parentheses:

000 (0)	001 (1)	010 (2)	011 (3)
100 (4)	101 (5)	110 (6)	111 (7)

Even more bits are required to represent numbers larger than 7. For example, to represent the numbers 0–255 we need eight bits[2]. The collection of bits is called a **word**. So, 011 is a 3-bit word, and 10101010 is and 8-bit word. The term **byte** is commonly used to indicate an eight-bit word. The rightmost bit is the *least significant bit* (LSB) and the leftmost bit is the *most significant bit* (MSB). The use of binary numbers goes beyond the representation of integer numbers. There are several schemes to 'encode' other information, including the sign (+ or −), the decimal point, letters of the alphabet, and so forth. Besides data, binary code can be used to write instructions about what to do with the data.

8.2 Analog and digital signals

As discussed in section 3.2 the basic procedure of measuring sound entails the conversion of the time variation of acoustic pressure into a time variation of an electrical voltage. The variation of the voltage with time is continuous, and there is a voltage value at any given time. The same applies to the acoustic pressure. These are **analog signals**. In the same way, the voltage from the output of an amplifier that sets the 'cone' of a loudspeaker into a vibration is an analog signal. Figure 8.1 shows a *triangular* waveform that has *continuous* values between 0 and 3 V, in other words, an analog signal.

If we want to record and store this analog signal, we have to take a series of voltage measurements at different time intervals. These data points are like 'snapshots' of the continuous variation of the voltage. To take a snapshot of the voltage, we need to stop the action long enough to allow the measuring device to acquire the voltage value. Once the device is finished with one measurement it is ready to take the next measurement, so in a way we missed what has happened in the time between the two measurements. As a result, we end up with a set of **discrete** values representing the continuous analog signal of figure 8.1. Suppose that we had time to take 7 measurements in total. Then our results will probably be something like 0, 1, 2, 3, 2, 1, 0 V. So, we have **sampled** (meaning *measured*) the signal 7 times, and we

---

[2] For further discussion see the section 8.11.1.

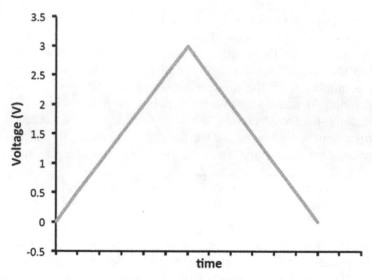

Figure 8.1. An analog signal.

have 7 **samples** (meaning *measurements*). If the measurements were taken at regular time intervals, and the whole process took 1 s, then we have a **sampling rate** of 7 Hz. If it took 2 s, then the sampling rate is 3.5 Hz. One way to 'summarize' the variation of the voltage is to use four discrete values. For instance, we could use the values 0, 1, 2, and 3. In this case we assign the value 0 V to any voltage lower than 0.5 V; we assign the value 1 V to any voltage between 0.5 and 1.5 V; we assign the value 2 V to any voltage between 1.5 and 2.5 V; and assign the value 3 V to any voltage above 2.5 V. In this way the triangular waveform is represented by four discrete values, as shown by the red graph in figure 8.2.

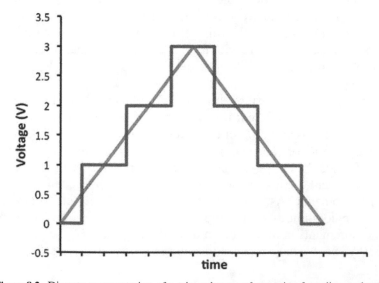

Figure 8.2. Discrete representation of a triangular waveform using four discrete levels.

This discrete representation (red line) in figure 8.2 deviates from the original triangular form (blue line) by 0.5 V at most, that is, we can have an error of ±0.5 V at most. The signal in this example is simple, and we have 4 different levels, 0 to 3V in decimal representation. We could write these numbers using 3-bit binary representation:

000 (0);   001 (1);   010 (2);   011(3)

Note that the MSB is always zero, so it is inefficient for representing numbers from 0 to 3. It is like writing a check for the amount of one dollar as $000 001. This would be a waste of time and ink. In terms of digital equipment, this would be a waste of speed and memory space. When processing a signal, it is common to use a number of levels that is a power of 2 (like 4, 8, 16 and so on) depending on number of bits we plan to use.

For the sake of argument, in this example, we started with a known signal. As discussed in section 3.2, for a real sound signal we need to take as many data points as required to ensure that we are not missing any 'wiggles' or other features that may be present. Essentially what happened here is that we used four discrete states. So, with our choice of levels (0, 1, 2, and 3), if the acoustic signal oscillates between 0 V and 0.4 V 10 times, or 10 000 times, we would have called all these values '0 V.' To get the detail needed to capture these variations of the signal, we need to use more levels. For instance, we may choose 10, or 50 or 256 levels and so on. What we are doing here is called **quantization**; we are quantizing an analog signal.

If we plan on using a computer to process this signal, we need to convert these levels into binary representation first. So, to represent the levels used in figure 8.2, we can use 2-bit binary words. We could quantize the signal using 256 levels. This would give us more detail, as shown in figure 8.3. Here we have more **resolution** compared to figure 8.2. Recall from section 8.1 that an eight-bit binary word can represent numbers from 0 to 255, so it allows 256 discrete levels. If we want to give this information to

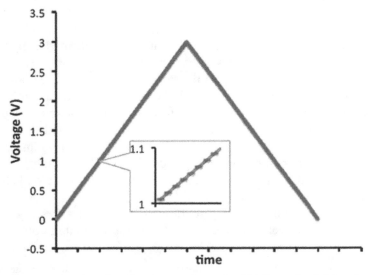

Figure 8.3. Discrete representation of a triangular waveform using 256 discrete levels. Note that the difference between the discrete and the analog values is much smaller than in figure 8.2.

a computer, we can use 8-bit binary numbers. By using 8-bits rather than 2-bits we diminish the error substantially, as shown in figure 8.3. The red plot is actually a number of discrete points. The inset shows a magnified view of the waveform around 1 V. Note that the deviation from the triangular waveform (blue line) is roughly ±0.006 V. In practice, the number of bits is usually from 8 to 32 bits.

To feed the information to a computer, we need a device that converts the quantized signal to binary numbers. We also need to have an agreement or protocol on how the computer should interpret these binary numbers. First of all, we need to know how many bits represent a number (in the example of figure 8.3 we used 8-bit numbers) and also the rate at which these numbers are coming. We also need to agree on how to read each binary number. For example, we can read from left to right. We need to tell the computer (or other digital device) that receives the binary information, or binary **code**, which device is sending the information, when to start reading data, when to stop reading data, and so on.

In summary, we started with an analog signal, then we quantized the signal using a number of levels, and then converted the numbers to binary representation. The next step is to convert this binary representation into a **digital signal**. In terms of audio applications, most of these steps are handled by sound cards, which are usually part of the computer.

8.3 Analog-to-digital and digital-to-analog conversion

The heart of a computer is the so-called central processing unit (CPU). The transistor is the basic element of the CPU and the basic element of most of the other components of the computer system. Depending on how it is configured, the transistor can act as an amplifier, for example, in sound equipment and portable 'transistor' radios. The transistor can also act as an electronic *switch*, and this is mainly how the transistor is used in computers. As a switch, the transistor can have two states: ON or OFF, and it can be used to switch a voltage, like switch on and off a voltage of 5 V. In this case the voltage can have two levels: 0 or 5 Volts. The two states of the transistor (ON or OFF) can also be indicated using the values 0 and 1, which are the values a binary digit can have[3].

Using a large number of transistors integrated in a very small area (about 7 billion in an area of about one square inch), the CPU can perform a huge number of logical and arithmetic operations in binary. Therefore, if we have a sound signal from a microphone (which is an analog signal) that needs to be processed by the computer, the analog signal must first be quantized, then converted to binary numbers and then to digital code that the computer understands. Once the computer has processed the digital signal, this digital signal must be converted back to analog before it is fed to an analog device, such as headphones.

There are two types of devices that are designed to do the necessary conversions: the analog-to-digital converter (ADC) and the digital-to-analog converter (DAC).

---

[3] These are nominal values. Usually any voltage between 2 and 5 V can be is used to indicate binary digit '1' and voltages between 0 and 0.8 V are used to indicate binary digit '0'.

The ADC converts all the incoming analog signals into binary code for further processing or storage by the computer. The DAC coverts the binary code and outputs the information as a quantized analog signal.

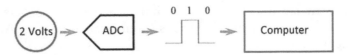

Figure 8.4. Path of a signal from an analog device (red circle) to a computer.

Figure 8.4 is an oversimplified diagram showing the path of an incoming analog signal. For example, suppose that we have a 3-bit ADC and that the signal level at this point in time is 2 V, which in 3-bit binary is represented by 010 (see list in section 8.1). The voltage is applied to the input side of the ADC, which converts it to its digital form, that is, 010. This number will appear at the output side of the ADC, in this case as a train of voltage pulses. This train of pulses is then routed to the CPU of the computer. During this time, the analog signal at the input may have changed, say to 3 V. Once the processing of the 2 V signal is complete, the ADC digitizes the next value, that is, the 3 V, and so on.

Figure 8.5. Path of a digital signal from a computer to an analog device.

Figure 8.5 shows the path of an outgoing digital signal from a computer. For example, suppose that the CPU completed an operation, and the result is 101. The decimal value of this binary number is 5 (see list in section 8.1). The digital signal is applied at the input side of the DAC in the form of voltage pulses. The DAC converts the digital signal to a quantized analog signal, and the result (5 V) appears at the output side of the DAC. This signal is now ready for use by an analog device, like headphones, and the DAC is ready to process the next digital number from the computer.

In many cases the ADC and DAC are parts of the same unit. The **sound card** on computers has several input and output ports. Figure 8.6 shows a computer sound

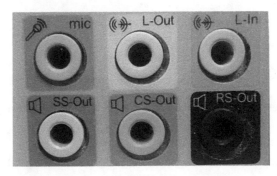

Figure 8.6. Computer sound card. All the signals going in and coming out are analog signal.

card. Typically, a sound card will have an input for low voltage signals (like the signal from a microphone, which is in the order of a few thousandths of one V); this input is pink. The light blue port is for input of the so-called line level signals, like signals from a radio or a tape player, which are in the order of 1 V. The lime green output is intended for headphones or as input to stereo amplifiers. The silver, orange, and black outputs are intended for surround-sound systems. All these input and output signals are analog signals.

8.4 Digital interfaces

One of the most significant aspects of new digital technologies is **connectivity**, which is their ability to communicate instructions and data. This is particularly true for music, including composition, performance and production. A very simple situation is a computer keyboard. Here the user hits a sequence of keys that have letters or symbols that are understandable to the user, for example, a sequence of backspaces to delete a word. Now this sequence must be sent to different parts of the computer (in the case of our example, to the screen and the document file), and in a form that is understandable to each of these parts of the computer. If more devices are connected, for example a printer, a cellphone, and the like, we need to make sure that the information intended for the printer goes to the printer, and not to the cellphone. So, what we need here are ways of directing the traffic of information between the computer and the devices connected to it. This is what we call a computer **interface**, which is a system that ensures that the outgoing information is in the correct form for the intended receiver, and that the incoming information is what the device expects to receive, and has the correct form.

There are several options for exchanging information in binary. For example, suppose that we have a computer, and we need to send the 4-bit word:

to a printer. Recall that the 0's and 1's represent nominal voltages of 0 and 5 V respectively. So, one way to communicate the data is to have four wires, and one shared ground wire, that is, five in total, as shown in figure 8.7, and the proper connectors at the ends. These connectors should have a minimum of five pins[4].

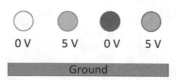

Figure 8.7. Schematic of pins and signal in a 4-bit parallel interface. Binary word is 0101.

---

[4] For a 8-bit data, the total number of pins can be a few dozen. The reason for the additional pins will become clear later in this chapter.

Once this number is sent to the printer, the computer must send the next number, for example:

So, the computer must change the voltages in figure 8.7 to 0,5,5,5, and so on for the subsequent words to be transmitted. Note that the digits are transmitted simultaneously, over separate wires in the cable. This is called a **parallel connection** or a **parallel interface**.

Another way to send the number 0101 to the printer is to send the digital bits one by one, using one wire for the signal and one for ground. This is shown in figure 8.8.

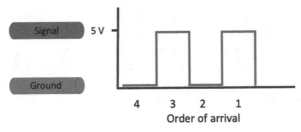

Figure 8.8. Schematic of pins and signal in a 4-bit serial interface. Binary word is 0101. In this diagram the LSB is received first, which is usually the case.

Once this number is sent to the printer, the computer sends the next four-bit word, and so on. In this case the bits are transmitted sequentially and we have a so-called **serial connection** or **serial interface**, which requires a *minimum* of two wires.

8.5 Bit rate and byte rate

The speed at which the bits are transmitted is called **bitrate**. In serial transmission, the bitrate is essentially determined by the duration of each pulse. For example, if the duration of each bit in figure 8.8 is 1/4 of a second, then the computer can send no more than 4 bits per second, or 4 **bps**, for short. In practice, this rate can be in the thousands (**kbps**) or millions (**Mbps**) and billions (**Gbps**).

As mentioned in section 8.1, an 8-bit word is called a byte. It is common to use the byte rate, measured in **bytes per second**, **Bps** for short. In practice we have **kBps**, **MBps**, and **GBps**. As one byte has 8 bits, it follows that 1 Bps = 8 bps, 1kBps = 8 kbps, and so on[5].

8.6 Communication protocol

In digital communication, a binary word can be used to represent very different things. For example, the 8-bit word:

---

[5] In practice, 12, 16, 24, 32-bit words are widely used.

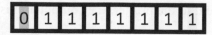

is the number 127 in decimal. But depending of the 'context,' in computer communications it can be used to indicate the command 'delete previous character' or 'the pitch of this note G9.' So, which one is it? It is clear that we need additional bits that contain instructions on how to interpret the binary numbers. These instructions should include information about the length of the words. Is it 8, 12, 16-bit? What is the bitrate? Which way to read the signal? That is, should we send/read the binary words starting from the LSB (yellow) or the MSB (green)? What is the meaning of the signal? Is it a value of some parameter, like setting the pitch of a note? The complete set of instructions is what we call the communication **protocol** for the specific interface. The data can be transmitted one bit or byte at a time, in which case we have **asynchronous** transmission. The rate of data transmission in this case is not steady. The data can also be in larger packages, and at a constant rate. This is called **synchronous** transmission, which is faster.

In serial communication, a good portion of this information is contained as additional bits to the data. For instance, the 8-bit data, like the one in the beginning of this section, may have an additional leading bit, and one or more trailing bits to indicate the beginning and the end of the word, and additional bits to indicate the source and destination of the digital message. So, what is exchanged between communicating devices is a *frame* or *packet* of bits containing the actual data (the *payload*) and the additional bits (the *overhead*). The overhead bits can be a significant percentage of the total packet. In terms of speed, this means that a 12 Mbps, may actually transmit data at 10 Mbps, and the remaining 2 Mbps are for the overhead.

For a parallel interface, this additional information means additional wires in the cable, and more pins on the connectors. The larger number of wires make parallel interfacing inherently faster than serial interfacing. But this also means that the cables are thick, and the connectors bulky. This is the main reason why parallel interfacing is mostly limited to communication within the computer and the communication between devices uses serial connections.

8.7 Hardwired transmission of digital data

A computer has several interfaces that control the exchange of information with other devices. From the computer interfaces, the electrical signals carrying the instructions and data in the form required by the communication protocol can go to the intended device through *wires*. In such a case we have a **hardwired connection**. The information can also go through radio waves, in which case we have a **wireless connection**. Wireless connections will be discussed in the next section.

There are two types of hardwired connections in use, one involves electrical signals traveling through metal wires, and the other involves light signals traveling though optical fibers. In this section we will overview two digital serial interfaces of interest to music, which are the USB and MIDI interfaces. The data transmission in USB is synchronous, and in MIDI is asynchronous.

8.7.1 USB interface

In the **USB** (Universal Serial Bus) one device (computer) is the master, which connects to the peripheral devices (printer, music keyboard, and the like). The *connectors* are labelled A, B, C, and the USB *versions* 1, 2, 3, 4[6]. The number of pins and the shape of the connectors have changed significantly over the years. The USB 2.0 uses 4-pin connectors[7]. For the USB 3.0 version, the A and B connectors have 9 pins. The newer type-C connector has 24 pins[8]. The bit rate continues to increase with each version. From 12 Mbps for the USB 1.1, to 480 Mbps for USB 2.0, to 20 Gbps for the more recent versions of USB 3.2, and even higher for the USB 4. Computers usually have two USB ports, which is generally inadequate for music applications. To solve this problem, one can use USB *hubs*, and configure 127 devices into a USB network. The host assigns to each peripheral device a different number (or *channel* as it is called) from 1 to 127. Each packet from the host contains the channel number of the intended receiver. Each peripheral ignores all packets, except the one containing its own channel number. When the host recognizes a new peripheral, it *temporarily* assigns it channel 0. So, the total number of channels is 128, that is 0 to 127.

Like all hardwired connections, the quality of the transmission is affected by the length of the cable, and there are several good reasons for limiting the length. For one, there is always potential for electrical interference from other cables or circuits, and the longer the cable, the more likely it is to pick up some interference which may alter the bits in the digital signal. But also, longer cables have higher resistance and higher capacitance (see section 7.2) that tend to reduce the strength of the signal, for instance a pulse of 5 V (that is the '1' in digital) may be reduced to 0.5 V which is the '0' in digital. Besides reducing the voltage of the pulses, for rapidly changing signals, like the fast succession of pulses in the high-speed digital signal, the impedance due to the higher capacitance can also make the pulses wide enough to overlap each other, so the signal becomes unreadable. Another factor is the time it takes for the signal to travel the length of the wire. For a 1-meter wire, it takes 5 billionths of a second (5 ns)[9]. A few billionths of a second may sound like an extremely short time to worry about, but remember that we are dealing with up to 20 billion pulses per second in USB 3.2, meaning that the duration of the pulse is already *under* one billionth of a second. The recommended length for USB 2.0 is 5 meters (about 16 feet) and 1 meter (about 3.3 feet) for USB 3.2.

Another source of problems with hardwired transmission comes from connecting wires. At each connection point we have a change in impedance (see section 2.4.3)

---

[6] USB 1 has been totally replaced by USB 2.

[7] Two for the digital signal, indicated by D+ and D-, one ground (GND), and one (Vbus) carries about 5 V and is used to power peripheral devices, such as a keyboard. The digital signal is between GND and the difference between D+ and D−. Using a so-called differential pair (D+ and D−) rather than one wire, is a common way to protect the signal from interference. See also section 7.8.2.

[8] There is no *up versus down* side in type-C, meaning that the effective number of pins is 12. Also, there is some redundancy within the 12 pins. USB 3.1 uses 9 wires.

[9] A voltage pulse travels through a pure copper wire at a speed of about 0.2 billion meters per second.

which means that at each connection point, some of the signal is reflected back, and interferes with the incoming signal. As a general rule, using cable extensions is not a good practice. In the case of USB, an *active or repeater* cable is required. Otherwise, even if one could find the correct combination of connectors/adaptors to make a five-meter USB cable out of five one-meter sections, the combination will probably fail to transmit a readable signal.

8.7.2 MIDI interface

MIDI is an abbreviation for Musical Instrument Digital Interface, and as the name suggests it is designed for communication between digital devices, as discussed in section 12.5. In its simplest version, each MIDI device has a MIDI IN port, to receive data from other MIDI devices and a MIDI OUT port, to send data. Many devices have the MIDI THRU port used to forward the signal received at the IN port, to the IN port of other MIDI devices in the network. All ports look the same, and the connections use the so-called 5-pin DIN connector, shown in figure 8.9.

The MIDI protocol uses 8-bit words, plus two additional bits: 1 bit (the start bit) to indicate the beginning and 1 bit (the stop bit) to indicate the end of the packet, that is 10 bits in total. The bit rate is 31.25 kbps. Just like USB, the MIDI interface has channels, 16 in total. Meaning that we can have 16 different devices connected at once, one per channel. Unlike USB, in the MIDI protocol, each device has its own number, and that number is set by the factory and can be reset by the user to avoid duplicate numbers in the connected devices. Each device accepts packages labeled with its own channel number, and ignores the rest. The first major revision, MIDI 2, was announced in 2020, and will be eventually introduced in newer machines. One of the significant changes is the increase in the size of the data packages, and increased bitrate. The benefit of increased bitrate is self-explanatory, but to do so will require a different interface, and indeed the shift seems to be towards the USB connection. To illustrate the advantage of using more bits, suppose that we are using MIDI to send instructions to an electronic keyboard. There are many different ways of hitting one note, starting from very soft to very hard. In music these differences are indicated by instructions like *pianissimo* and *fortissimo*, and the like[10].

Figure 8.9. MIDI ports IN and OUT. The digital signal is between top right pin, and top middle pin.

---

[10] This has to do with the dynamic range discussed in section 7.6.

MIDI refers to this difference as *velocity* and uses 7 bits to specify the velocity value. This means that there are $2^7 = 128$ distinct levels. So, the range between the loudest a note can be played and the softest that can be played is divided into 127 levels. Using 16 bits, as is planned in MIDI 2, will give more bit-depth, and will increase the dynamic range by a factor of over 500.

8.8 Wireless transmission of digital data

8.8.1 WiFi wireless connection

A typical home or office internet setup may use both hardwired and wireless connections. The first step is to bring in (the house or office) the *internet signal from an internet service provider* (ISP). In most cases this happens through the so-called Ethernet, via a cable or a fiber optic, which attaches directly to the user's **modem**. The **router** (which may be in the same box with the modem) distributes the digital signal to all of the user's devices. The devices (computers, printers, loudspeakers, and the like) may be physically hardwired to the router, *or* may be connected to the router by radio wave transmission. This is the so-called **WiFi**, which is an interface that talks and listens to all the devices connected to it, and allows communication *between* all these devices (laptop, printer, cell phone, and so on). The digital signal (instructions and data) is riding on a carrier wave (or more properly, *the digital signal modulates the carrier wave*) in a way similar to FM radio, discussed in section 1.6. The carrier frequencies used are most commonly in the 2.4 to 2.5 GHz band, the so-called 2G, or in the 5 to 6 GHz band, the so-called 5G[11]. Both 2G and 5G are in the so-called *microwave* part of the electromagnetic spectrum (see section 1.6). The modem, router, cables, and WiFi together form the Local Area Network, LAN for short.

8.8.2 Bluetooth wireless connection

Another popular wireless network is the **Bluetooth**, which is used to exchange *digital* information between devices without involving WiFi or the internet. This type of connection makes a *local* network of devices, a so-called Personal Area Network (PAN for short). Again, the devices in the network can talk and listen to each other, for example, a cell phone can send data (music) and instructions (change volume, reduce bass, and the like) to other devices such as earphones, loudspeakers, amplifiers, and other. Bluetooth operates in the 2.4 to 2.5 GHz frequency band (same as 2G in WiFi) and roughly in the same way. As in USB, we have different channels, a total of 79 in Bluetooth. Each channel has frequency bandwidth of 1 MHz[12]. In terms of sound, Bluetooth offers the big advantage of connecting wireless microphones, amplifiers, headphones, loudspeakers, and the like, to a computer (or cell phone or tablet), without any wires, as long as the devices are compatible with the computer. This compatibility, can be a problem at two levels. The *worse* situation is when the device is not a proper Bluetooth device, in which

---

[11] Which has nothing to do with the 5G mobile standard.
[12] Bluetooth Low Energy has 40 channels. The frequency bandwidth of each channel is 2 MHz in this case.

case, the computer will not recognize the device. On a different level, we need to keep in mind that the data bits can be combined in different ways to represent the digital information, and also, the bits can be sent at different rates. For example, we can send/receive our digital information in packs of 16 or 24 bits. Depending on the application, the number of bits used may have a significant effect on the quality of the signal. For example, the 24-bit scheme allows better sound quality. We could also send the bits at different rates, which affects the size of files that can be transmitted. So, there are different possibilities, the so-called Bluetooth **codecs**, with significantly different levels of performance. These codecs are not supported by all computers. If the codec of the device is not supported by the computer, the connection will go to the default codec, and the resulting sound quality may be lower than expected from the device's specifications and price.

8.8.3 Potential problems with wireless transmission

Both WiFi and Bluetooth have some resemblance to ordinary radio, in that they involve a transmitter and a receiver. Both transmitter and receiver have antennas, and the signal is carried by a high frequency carrier wave. Once the modulated wave (carrier plus signal) is picked up by the receiver's antenna[13], the carrier wave is separated out, while the signal, which is digital in both WiFi and Bluetooth, continues for further processing by the receiving device. The process of combining carrier wave and signal, and then separating the signal from the carrier takes some time, which can be in the order of tenths of a second. This **lag** or **latency** can be a problem in some applications, and can be noticeable at times, for example when combining audio and video.

The range (typically a few tens of meters) of WiFi is generally higher than Bluetooth (usually under 10 m). The range is higher when there is a clear line of sight between devices, although both can transmit through walls and around furniture. The range can be increased by increasing the power output of the emitter *and* the connected devices (cell phones, printers, earbuds, and so on). One of the problems in doing so is that the frequencies used in WiFi and Bluetooth are within the range of 1 to 30 GHz which is the **microwave range**, and there is public concern about the potential health effects from increased exposure to this radiation[14].

The power carried by the radio waves of both WiFi and Bluetooth (as well as analog wireless connections) is low, generally under 0.1 Watt. In practice, this means that all the interconnected devices must have their own power source. This is important to keep in mind, especially when planning networks that involve Bluetooth loudspeakers. Neither the WiFi nor the Bluetooth signals are strong enough to drive a loudspeaker. This means that an amplifier is required, to receive the WiFi or Bluetooth signal, separate out the audio part and amplify the audio signal to the level required to drive the loudspeaker, which is a few Watts to several tens of Watts. The amplifier is in the enclosure of the loudspeaker. If more than one

---

[13] The antennas are typically in the order of a few centimetres and may not be visible.

[14] Ordinary microwave ovens operate at 2.45 GHz.

loudspeaker is used, the additional loudspeakers need not have an amplifier. In this case, each additional loudspeaker must be *hardwired* to the loudspeaker that has the amplifier.

8.8.4 Why use microwaves for wireless?

One may wonder about the need to use high carrier frequencies. The major advantage in using frequencies in the microwave range is the size of the required antenna. Here there is some analogy with the standing waves in pipes, discussed in chapter 6, because there is a relation between the length of the pipe and the allowed modes. As it turns out, the length of the antenna should be about 1/4 of the wavelength of the carrier signal. If the frequency, f, of the carrier radio wave is 3 MHz, and c is the speed of light, then the wavelength is $c/f = 100$ m. This means that the antenna's length must be $100/4 = 25$ m, which is over 80 ft. If the carrier frequency is 3 GHz, the wavelength is 0.1 m, which requires an antenna 0.025 m long (just under one inch) which can fit easily in most peripheral devices.

8.8.5 Wireless transmission of analog signals

Besides digital signals, wireless connectivity is also very useful for analog signals. Wireless microphones and various instrument pickups (see section 11.6.1) are very popular because they allow performers to move around the stage. The principle is essentially the same as with Bluetooth, the range of carrier frequencies used is similar, and the same distance limitations apply.

8.9 Bandwidth

The term bandwidth may have different meanings. In digital data transmission, such as Ethernet and WiFi, it is the transmission rate measured in bits per second (bps, see section 8.5) and of course more bandwidth is better, because it means faster communication. In analog circuitry and analog devices, bandwidth means a range of frequencies, for example, the bandwidth of an amplifier, discussed in section 7.6. Recall that the frequency range of audible sound is from 20 Hz to 20 kHz, so the 'bandwidth' is about 20 kHz.

In AM radio (section 1.6) we use the audio signal to modulate the amplitude of a radio frequency carrier wave. The resulting wave is no-longer a simple sinusoid, because the amplitude is changing. This means that we have additional frequencies, centered around the frequency of the carrier wave. Suppose for example that the frequency of the carrier wave is 700 kHz. If we modulate the amplitude of the carrier wave with a frequency of 1 kHz, the spectrum of the resulting wave will have components at frequencies 699 and 701 kHz[15]. If we modulate with a signal of 5 kHz, we will have components at 695 and 705 kHz. So, each modulating frequency adds two side-bands centered around the carrier frequency, as shown in figure 8.10.

---

[15] The interested reader can show this using essentially the same math applied in appendix B.3.

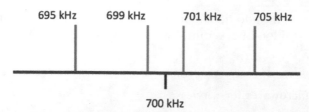

Figure 8.10. Side bands of 700 kHz carrier signal. From 1 kHz modulation (blue) and 5 kHz modulation (red).

It follows then that if we use frequencies representing the entire audible spectrum, we will have sidebands extending from 680 to 720 kHz, and a bandwidth of 40 kHz, which is two times the highest frequency of the modulating signal, that is $2 \times 20 = 40$ kHz for the entire audible range. Actually, to make room for more stations[16], each AM radio station is allowed a bandwidth of 10 kHz, so in AM radio we can only transmit sound frequencies up to 5 kHz! And cell phone bandwidths are even lower, about 4 kHz. These bandwidths are adequate for voice (see section 12.4). This is why music sounds better on FM, because FM channels are allowed 200 kHz bands, which is enough to accommodate the entire audible spectrum, and enough room for accommodating the two channels required for stereo sound[17]. The situation is more or less the same for wireless transmission of digital data. WiFi 2G accommodates 11 channels and the bandwidth of each channel is 22 MHz. Note that $11 \times 22 = 242$ MHz, and that the frequency range of 2G is only 2.4–$2.5 = 100$ MHz. What this means is that *some* channels overlap each other which may cause interference. Here, the general conclusion is that more bandwidth per channel allows more detail/better quality/more information. On the other hand, more bandwidth per channel, means a smaller number of channels.

8.10 Further discussion

Minimum sampling rate

In measuring or in digitizing a signal, it is important to know the rate at which we should be taking measurements, the so-called sampling rate discussed in section 8.2. As discussed in section 3.2, the sampling rate depends on the frequency components present in the analog signal. Here we will use a simple example to illustrate the significance of the sampling rate. Suppose that we have a pure tone of frequency 660 Hz. This means that it takes $1/660 = 1.515$ milliseconds to complete one cycle. Now suppose that we *sample* the signal every 1.6 ms, in other words, take a measurement of the signal every 0.0016 s. One sample every 0.0016 s means a *sampling rate* of $1/0.0016 = 625$ Hz. The result of sampling at this rate is shown by the black dots in figure 8.11, which is pure sinusoidal of frequency equal to 35 Hz and *has nothing to do with our 660 Hz signal*. So, the low sampling rate produces **aliasing**, meaning a waveform with frequency that relates to a combination of the true signal frequency

---

[16] The AM range can fit 106 stations in total.
[17] The FM range can fit 100 stations.

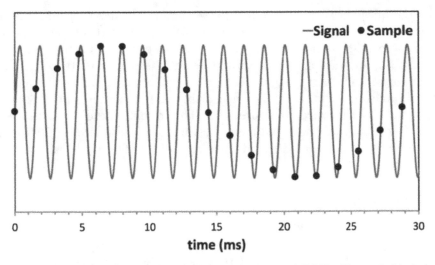

Figure 8.11. A signal of frequency 660 Hz, sampled at a frequency of 625 Hz. The result (black dots) is a sinusoidal waveform of frequency 660–625 = 35 Hz.

and the sampling frequency (660–625 = 35 Hz in this case). If we change the sampling frequency to 615 Hz, we get an alias of frequency 45 Hz!

The problem comes from the fact that we have less than one point inside a full cycle of the signal, and that is not enough. The absolute minimum would be two points within the cycle, for example, the positive peak and the negative peak. With this information we would know how to place the sinusoid, and of course, more points would make it easier to recognize. This is done in figure 8.12. Here we sample every 0.3030 ms, that is a sampling rate of 3300 Hz, and the black dots (five within each cycle) outline a sinusoidal of the same frequency and amplitude as our signal. In practice, the problem is slightly different in that we have a signal containing many frequencies.

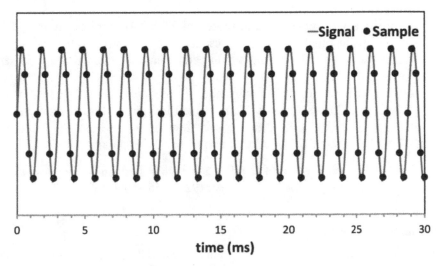

Figure 8.12. A signal of frequency 660 Hz, sampled at a frequency of 3300 Hz. The black dots show a recognizable sinusoidal waveform of frequency 660 Hz.

Now the question is: how fast should we sample the signal? From our simple example above, we know that we need to sample at a rate that is higher than any of the frequencies in the sample so that we have at least two points within each cycle of the signal. So, the sampling rate should be at least equal to twice the highest frequency present in the sample. Otherwise, we will not catch all the frequencies in the signal, and might end up with aliasing, that is, frequencies that are not in the actual signal, as happened in figure 8.11. This result is known as the **Nyquist theorem**. For sound, we know that the highest frequency of interest is 20 000 Hz, so we should be sampling sound signals at a rate of 40 000 Hz. Most sound cards sample at 44 000 Hz.

8.11 Equations

8.11.1 Reading binary numbers

In the familiar decimal system, the base is 10, and the digits (0 through 9) are understood to multiply the powers of 10. For example, the 5-digit number 20 358 means

$$2 \times 10^4 + 0 \times 10^3 + 3 \times 10^2 + 5 \times 10^1 + 8 \times 10^0 = 2 \times 10\,000 + 3 \times 100$$
$$+ 5 \times 10 + 8 \times 1 = 20\,358.$$

In binary the base is 2, and the available digits are 0 and 1. The binary number 101101 means

$$1 \times 2^5 + 0 \times 2^4 + 1 \times 2^3 + 1 \times 2^2 + 0 \times 2^1 + 1 \times 2^0 = 1 \times 32 + 0 \times 16 + 1 \times 8$$
$$+ 1 \times 4 + 0 \times 2 + 1 \times 1$$
$$= 32 + 0 + 8 + 4 + 0 + 1 = 45.$$

With six digits in decimal, we can represent 10^6 (= 1 000 000) numbers, namely from 0 to 999 999. In general, using N digits we can represent 10^N numbers (from 0 to $10^N - 1$).

With six-bit binary numbers we can represent 2^6 (= 64) numbers, namely from 0 to 63. Using N bits in binary we can represent 2^N numbers (from 0 to $2^N - 1$).

It is common to use eight-bit binary numbers, meaning that we can represent 2^8 (= 256) numbers, from 0 to 255.

In computer science, it is also common to use **hexadecimal** (**hex** for short) numbers. The base in hex is 16, for which we need 16 digits. This is accomplished by using alphanumeric digits. In addition to the familiar 0–9, we use A through F to represent 10 to 15, respectively. In hex the number 210B means

$$2 \times 16^3 + 1 \times 16^2 + 0 \times 16^1 + 11 \times 16^0 = 2 \times 4096 + 1 \times 256 + 0 \times 16 + 11 \times 1$$
$$= 8192 + 256 + 0 + 11 = 8459.$$

8.11.2 Quantization error

The process of assigning discrete values (figures 8.2 and 8.3) to a continuous (analog) signal is called **quantization**. As discussed above, the discrete form deviated from the analog form, and this deviation, or **quantization error** depends on the number of discrete levels introduced, which in the binary form is determined by the number of bits used. If our signal varies from 0 to A ($= 3$ for example in figure 8.2) and we use N bits in our binary representation, then we have 2^N *levels*, or $2^N - 1$ *steps* in between 0 and A. The difference between the digital signal and the quantized signal can be half of a step at most. So, the quantization error can be at most $\pm 1/2\ A/(2^N - 1)$. In the example of figure 8.3, where we use eight bits (recall $2^8 = 256$), and $A = 3$, the quantization error is $\pm (1/2)\ 3/(256 - 1) = \pm 0.006$ V.

8.12 Problems and questions

1.   (a) Express the decimal number 6 in binary.
       (b) Express the binary number 0101 in decimal.
2. Is the voltage of the wall outlets analog or digital? Explain.
3. Refer to figure 8.6. Which of the input/output ports are digital?
4. Why do digital computers work with binary numbers?
5. In figure 8.2 we used 4 levels for quantization.
       (a) Was this a good choice if we plan to use 4-bit words?
       (b) Was this a good choice if we plan to use 2-bit words?
       (c) Was this a good choice if we plan to use 1-bit words?
6. Use an example to demonstrate the difference between digital signal and binary signal?
7.   (a) Cell phones use a bandwidth of about 4 kHz. What is the highest audio frequency in the phone signal?
       (b) Is this sufficient for listening to music over a phone conversation?
       (c) We can use a cell phone to play music with acceptable sound quality. How is that possible?

IOP Publishing

The Physics of Sound Waves (Second Edition)
Music, instruments, and sound equipment
Panos Photinos

Chapter 9

The musical environment

Royal Albert Hall. Attribution: Colin/Wikimedia/CC BY-SA-4.0.

9.1 Introduction

From ancient times, significant thought and effort has been devoted to improve the acoustics of temples, theaters, and other places of public gatherings. The scientific approach to improving acoustics dates back to the late 19th Century, and today, lecture rooms, concert halls, and other, benefit from well-established design principles from the science and engineering of sound. But because of the many parameters that need to be factored in, the designs are not always successful, at least initially. In addition, the quality of the sound in a hall may be best for one particular purpose, for example, an orchestra performance, but unsuitable for singing.

In this chapter, we will apply concepts introduced in earlier chapters to understand the physical quantities that affect the quality of sound, both outdoor and indoor. We discuss which criteria should be applied to make a particular space more suitable for a specific purpose, such as orchestral performances, speech, and so on. For instance, what criteria should be used to achieve better sound quality for a music room in a house versus a music hall? We will start with a brief review of the important processes involved in indoor and outdoor acoustics.

9.2 Review of the fundamental processes

The sound emitted from a sound source tends to spread in all directions (see section 2.7). As a result, the intensity of sound decreases as the distance between the source and the listener increases following the inverse square law. So, if we have one seat at an outdoor concert in the front, say at 10 m from the stage and one seat at 20 m (that is, if the distance is doubled) and along the same line of sight from the stage, the intensity at the back seat will be $1/2^2 = 1/4$ of the intensity received at the front seat. Also recall that high frequencies are attenuated by absorption in air more than low frequencies. In addition, low frequencies tend to spread sideways more than high frequencies.

From the reflection of sound (see section 2.4.1) we know that we can hear the echo of our voice reflected from a wall, if our distance to the wall is 17 m (about 50 feet) or more. So, the round-trip distance in this situation is $2 \times 17 = 34$ m (about 100 feet). As sound travels at a speed of about 343 m s^{-1}, the round-trip time is $34/343 = 0.1$ s, or 100 ms. This means that if there is any reflection from a wall that is closer than 17 m, the sound reflected from that wall will 'fuse' together with the original sound. Also, we will need to recall that when sound waves hit an obstacle, like a wall, the obstacle reflects part of the wave energy and absorbs part of the wave energy. The percentage of absorbed energy depends on the material the obstacle is made of and the surface texture of the obstacle. Softer materials (drapes, pillows, and so on) absorb more (and reflect less) than hard materials (such as concrete and hardwood floors).

9.3 Outdoor acoustics

The main concern in an outdoor environment is to minimize the loss due primarily to spreading and absorption. Figure 9.1 illustrates the path of sound waves in an outdoor setting. Note that grass mostly absorbs the sound energy and reflects very little of it. Also, waves moving upwards do not reach the listener, in other words, the usable sound energy is lost.

One way to remedy the situation is to add barriers behind, above, and to the sides of the stage, as shown in figure 9.2. One can use barriers with the right spacing from the sound source and the proper orientation and surface structure, to direct the sound energy towards the audience. In terms of distance the barriers should be close to sound sources on the stage. The lateral barriers should be placed symmetrically. When setting the reflecting barriers, one should keep in mind that the sound reaching the audience should be an even blend of all the sounds on the stage. Also, the sound

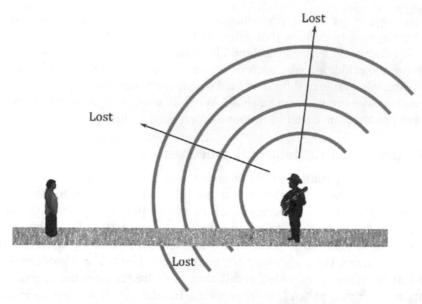

Figure 9.1. Path of sound waves in an outdoor setting.

Figure 9.2. A band shell diffuses the sound towards the audience.

from the stage should be distributed evenly in all directions of the line of sight from the stage to the audience. So, the term 'reflection' here actually means more of *diffuse reflection* rather than *specular reflection* (see section 2.4.1.)

Another problem, which can affect both indoor and outdoor acoustics, is that if the whole setting is flat, people standing in the front act as barriers that partially block the sound from reaching the audience in the back. One way to solve this problem is to lift the stage above ground. This allows the audience in the back to have an unobstructed line of sight to the stage, and receive more sound directly from the stage. For the same reason, if loudspeakers are used, they should be above ground. Note that in figure 9.2, the loudspeakers are mounted at the top of the band shell. Another solution is to have the audience area sloping upward from the stage, allowing unobstructed line of sight to the stage. This is the basic idea behind the *amphitheater* design common in modern and ancient outdoor venues.

The sound intensity depends on the distance of the listener from the source. This happens because the sound spreads out, and also because air absorbs some of the sound energy. In a large venue, the sound intensity level can drop significantly beyond the first few dozens of rows. In this case, the use of additional loudspeakers becomes necessary. More loudspeakers help the intensity level, but can also cause the quality of sound to deteriorate. To understand how this deterioration comes about, note that the electrical signal from the amplifiers gets to the loudspeakers almost instantly. A listener gets the sound from the nearest loudspeaker first, followed by the sound from the loudspeakers on the stage or other loudspeakers in the venue. These sounds travel at 343 m s^{-1} and depending on the distances, all these sounds may arrive with considerable delay. For a large venue, this delay can be 100 ms or more and the listener perceives one or more echoes which become a problem. One way to address this problem is to use an electronic compensation scheme to *delay* the signal going from the amplifier to the loudspeakers in the back, so that the sound arrives at the audience in the back approximately at the same time as the sound from the loudspeakers on the stage. Another echo related problem happens in outdoor venues if there are barriers that are just far enough to produce an audible echo, that is barriers at distances around 17 m (50 ft) or more. This is a common problem in events at school and college campuses. These echoes are detrimental to the sound quality, and this is why good outdoor venues are by design located in open spaces, surrounded by trees.

9.4 Indoor acoustics

As in outdoor venues, the basic objectives of large halls, and music listening rooms in a house, are to get a uniform blend of the sounds from the sources to the listeners and distribute the sound intensity in a uniform way as much as possible. Reflections that reinforce the sound are always present indoors by default. In an indoor space echo is not a problem. The reason is that there are many reflections, all arriving with very little time difference, much less than 0.1 s. The result is that the reflections fuse together. Figure 9.3 shows two such reflections, from the floor and the ceiling. The *direct sound* (blue arrow in figure 9.3) reaches the listener first, followed (in this example) by the reflection from the floor. This is the *first reflection*. The *second reflection* is the reflection from the ceiling, because it travels a slightly longer distance than the reflection from the floor, and arrives shortly thereafter. Many other reflections follow. For example, the sound may bounce from the back of the stage and reach the listener, or may bounce from the back of the stage to the ceiling and

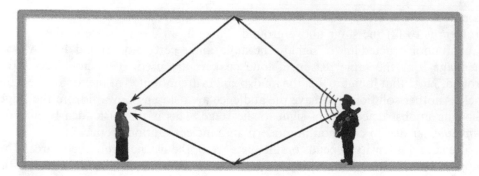

Figure 9.3. Direct sound (blue arrow) and reflected sound (black arrows) in a closed space.

then to the listener, or from the sidewalls, or the back of the hall, and so on. The order of arrival depends of the location of the listener. One problem that can occur from these reflections is that we may start building standing waves. For example, if the waves reflect directly from the back of the stage to the back wall of the room, we may have a situation similar to what happens in a pipe closed at both. As discussed in section 6.9, a number of vibrational modes will build up and resonate. The corresponding tones will sound louder, and the sound quality will deteriorate. The solution is to avoid flat walls, and add features that promote diffuse reflection, like the suspended white convex diffusers shown in the opening picture of this chapter.

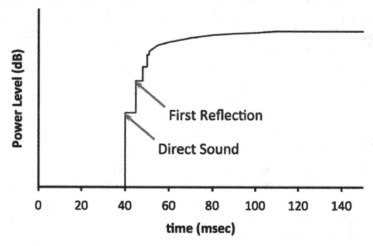

Figure 9.4. Power level for a steady long tone received at a typical point in a hall.

Figure 9.4 shows the evolution of sound received at a typical point in the audience area from a *steady long tone* played on the stage. The first step-like feature marks the arrival of the direct sound from the stage, followed by another step marking the arrival of the first reflection, and another step marking the arrival of the second reflection, and so on. As the subsequent reflections arrive with little time lag between them, they fuse together, and the power levels-off at the **reverberant level**, which is higher than the direct sound. It is higher because the sound source is a *steady long tone*, and we have sound reinforcement by the multitude of reflections.

The combination of direct and reflected sounds all together forms the **reverberant sound**. The sound power level remains at the reverberant level for as long as the steady tone is on. After the steady tone is turned off, the power level begins to die out. Figure 9.5 shows how the power level drops. Note that the decay starts at 50 ms. The time it takes for the reverberant sound to drop by 60 dB from the reverberant level (which means that the intensity drops to one millionth of the reverberant level) is called the **reverberation time**. In figure 9.5 the decay indicated by the black line corresponds to reverberation time equal to about 150–50 = 100 ms, and the red line corresponds to reverberation time of about 200–50 = 150 ms. Note that both lines are straight. This is a desirable quality for a music hall, indicating that the sound is blended evenly.

With each reflection, some sound energy is lost to absorption. The major part of absorption occurs at each reflection (by the walls, seats, furniture, audience, and so on). To a lesser extent, sound energy is lost because the air in the hall absorbs part of the energy, especially the higher frequencies. In a larger room, the distance between walls on average is longer and the sound waves travel for a longer time between reflections, compared to what happens in a smaller room. This means that in a larger room the sound dies out slower. This makes the reverberation time *longer* for *larger rooms*. Also, the reverberation time depends on the percentage of absorption at each reflection. If the walls do not absorb significantly, the sound waves will take many bounces before their energy is lost. In other words, the reverberation time is *longer if the absorption at the walls is low*, and vice versa. Of course, there are different reflecting surfaces in a room or a hall: walls, chairs, carpets, and so forth, which means different percentages of absorption at different parts of the hall. In addition, the presence of an audience affects the reverberation time as well. When empty, a hall will have longer reverberation time than when it is full of people. So, it is best to have seats that are made of soft cloth upholstery, which absorbs roughly as much as the human body. In this case the absorption is roughly the same whether there is an

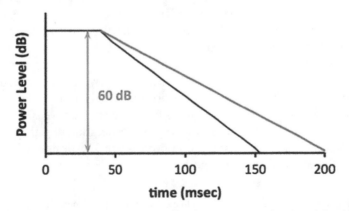

Figure 9.5. Reverberation time for the black line is 100 ms, and for the red line 150 ms.

audience or not. The reverberation time also depends on the frequency. Higher frequencies die out quicker than lower frequencies, due to absorption by the air in the room. The calculation of the reverberation time can be rather complicated. There are specialized room acoustics software that can calculate the reverberation time from the dimensions and the wall materials[1].

9.5 Sound qualities of halls

When describing the sound quality of large halls, terms such as 'fullness', 'intimacy', and the like, are commonly used. Here we will try to relate these subjective terms to measurable quantities. First of all, **liveness** is directly related to the reverberation time. Long reverberation time gives liveness to the hall. Depending on the purpose of the space, a long or short reverberation time may be more suitable. For example, in a lecture hall, reverberation times should be short, under 0.5 s. Longer reverberation times have a negative impact on the clarity of speech. On the other end of the spectrum, to really enjoy organ music, a much longer reverberation time (about 2 s) is more suitable. Symphony halls have reverberation times in the range of 1.5–2.5 s.

Intimacy is used to indicate the feeling of being close to the performers. This occurs when the first reflection arrives less than about 20 ms after the direct sound. In figure 9.4 the first reflection arrives 5 ms after the direct sound, which gives intimacy to the hall represented by figure 9.4. One way to achieve intimacy is to have a reflective structure above the stage. Adding reflecting surfaces on the sides helps as well, and in addition it helps blend the sound. **Blend** is important for large stages, otherwise at some spots in the hall some voices/instruments may sound louder/weaker than intended. For instance, seats towards the front right of a hall will receive more sound from the right side of the stage than they receive from the left side of the stage. This difference is larger with wider stages. The problem can be corrected to some extent using closely spaced lateral barriers on each side of the stage.

If the direct sound is a small portion of the reverberant sound level then the sound we perceive comes mostly from the reinforcing reflections. A long reverberation time combined with a reverberant level that is much higher than the direct sound result in what is known as **fullness**. The *opposite* of fullness is **clarity**, which results when the reverberation time is short and the reverberant level is not much larger than the direct sound (that is, when the reinforcement from the reflections is weak). As we cannot have high clarity and fullness together, some compromise is necessary.

If the reverberation time for low frequencies is longer than the reverberation time for high frequencies then we speak of **warmth**. If this is not the case, and the reverberation time is about the same for all frequencies, then we speak of **brilliance**. Again, these two characteristics are in a way opposite, meaning that we can't have *both to the maximum* and that some compromise in the design will be necessary depending on the main purpose of the hall.

---

[1] For further discussion see section 9.12. A free online calculator can be found at: https://www.atsacoustics.com/page–Free-Online-Room-Acoustics-Analysis–ora.html.

9.6 Sound qualities of small rooms

The walls in most rooms in homes are very close to the listener, which means that the first reflection arrives well under 20 ms after the direct sound. So, intimacy is always there by default. If the average dimension of the room is 5 m (15 ft) then sound bounces off the walls over 70 times in each second, which means the effect of absorption will be very significant compared to a large hall. So, because of their small size, and the relatively high absorption on surfaces usually found in a room (carpets, drapes, and the like), the reverberation time is typically less than 0.5 s. Short reverberation time means that liveness is always lacking in home listening rooms. The multitude of reflections in a small room reinforce the sound, in other words the reverberant level is higher than the direct sound. But as the reverberation time is so short (remember that the second requirement for fullness is long reverberation time) home listening rooms lack fullness. This means that in terms of fullness, all there is essentially, is what was captured on the medium, the CD for example, during the recording of the album.

9.7 High fidelity sound

To enjoy high quality recorded or transmitted sound, the reproduced sound must correctly include all the frequencies present in the original performance, without noise or distortion of the loudness, or compression of the range of loudness[2]. All these requirements are met by most home sound systems. In addition, the reproduced sound should capture the *reverberation characteristics* and the *spatial pattern* of the original sound. For example, if the original performance had the violins on the left and the cellos on the right of the audience, then the listener at home should also get the sensation of this spatial arrangement of the instruments. A system that meets all these requirements is a **high fidelity (HiFi)** system. **Stereophonic** and **surround sound** systems meet the last two requirements with varying degrees of success.

9.8 Stereophonic sound

Our binaural hearing (see section 4.12) allows us to identify the direction of the sound source. Our brain uses both the difference in arrival time and the difference in loudness to localize the sound source. As discussed above, the audience in a concert hall receives the direct sound first followed by the reverberant sound, which comes from *all* directions and in some cases can be *louder* than the direct sound. Under these conditions the audience can still identify the direction of the sound as coming from the orchestra because of the so-called **precedence effect.** The precedence effect tells us that our hearing is tuned to identify the direction of the *first* arriving signal as the direction of the source of the sound. In a music hall, that would be the direct sound, which always comes from the stage.

On the other hand, in a small room setting, if two identical sounds arrive from two different directions within 2 ms or less, then the source is perceived as being

---

[2] The range of loudness, or dynamic range, is discussed in section 7.6.

somewhere in between the two actual sources of the sound. If the intensity of one of the two sounds is higher, then the source is perceived as being somewhere in between the two sources, and closer to the side of the source with higher intensity. A simple calculation will show that if the distance from the listener to the loudspeaker changes by 1 m (about 3 ft) the arrival time changes by about 3 ms. This means that the 'in-between' effect is sensitive to the distance of the listener to the sources. So, a lateral shift of 1 m may be enough for the precedence effect to take over, in which case the direction is determined by the order of arrival. So, if the listener is closer to one of the loudspeakers in the room, then the listener will perceive this loudspeaker as the one source of the sound. From this example one can appreciate that the placement of the loudspeakers relative to the listener is very important in reproducing the spatial arrangement of the instruments during the recording.

In the simplest form, a stereo recording reproduction starts by placing two microphones: one picking up mostly the sounds from the left side of the stage (L) and one picking up mostly the sounds from the right side (R) of the stage, as shown in figure 9.6(a). Both microphones pick up the center stage, just about equally. The sound picked up by each microphone is recorded on a separate channel, L and R, respectively. On playback, the sound from each recorded channel is directed to a separate loudspeaker, L and R.

Of course, the sound from each loudspeaker will bounce around the room, and the reflections will be coming to the listener from everywhere. However, *if the listener is seated midway between the two loudspeakers*, the direct sound from

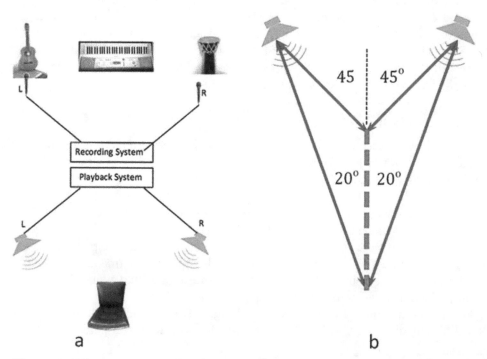

Figure 9.6. (a) Stereo recording and reproduction. (b) Placement of loudspeakers for stereo systems.

loudspeaker L will come to the listener's left ear first and at higher intensity. So, the listener will identify loudspeaker L as the source of what comes to the left ear, meaning that what was recorded by microphone L will appear as coming from the left. In figure 9.6(a) that would be the guitar. Similarly, the sound recorded by microphone R will appear as coming from the speaker on the right. In figure 9.6(a) that would be the sound of the drum. The sounds that were picked up equally by microphones L and R (the keyboard in figure 9.6(a)) and reproduced equally by loudspeakers L and R. So, if the listener is seated midway between L and R, the sound of the keyboard will appear as coming from the middle.

9.9 Placement of loudspeakers

Stereo sound relies on the arrival time and intensity of the signal received from the two channels; this makes the configuration of the listening space very important. Figure 9.6(b) shows a typical arrangement, where the speakers are placed near the corners of the room, and along the shorter wall. One reason for placing the speakers near the corners is that the sound is reinforced by reflection from both walls forming the corner. This is so because the wall acts like a closed end in a pipe, in other words, there is a pressure antinode at the wall (see section 6.7.) The reason for using the shorter wall is that it allows for more 'good listening space', the so-called **sweet spots**. This can be understood by examining the angles shown in figure 9.6(b). For best results, the listener should be seated on the center line between the speakers, not too close and not too far from the loud speakers, as indicated by the green dashed line. The key point here is that the loudspeakers should be at an angle of about 20–45° to the right and left of the listener. If the angle is too small, that is, if the speakers are too close to each other, the stereo effect is diminished. We essentially hear one source, that is a **monophonic** sound. If the angle is too large, we perceive two different sources with a 'hole' in the middle. In figure 9.6(b) the red solid lines indicate the 20° angles, and the blue lines are the 45° angles. The listener can be seated anywhere along the green line and have good stereo effect. If we assume that the speakers are set 4 m apart (12 ft) then the distance of the listener to the speakers must be 2.8 m (8.5 ft) for the 45° angles, and 5.8 m (17 ft) for the 20° angles. If the speakers were set 6 m (18 ft, which implies a wider wall) then the distances of the speakers to the listener should be 4.2 m (12.7 ft) and 8.5 m (26 ft) for 45° and 20° angles, respectively. These distances are too large for an ordinary house, and leave little room for placing the furniture without obstructing the line of sight from the listener to the speakers.

As the source localization depends on the intensity and time of arrival, one can use the **balance** control in a stereo system to adjust the sweet spot. In other words, the listener can sit to the right or left of the green center line, and increase the loudness of the left or right channel, respectively, and still have good stereophonic sound. The balance adjustment works as long as the difference in arrival times remains *under* 2 ms. In ordinary rooms, this difference in arrival times translates into a lateral displacement of roughly 2 m (6 ft) to the right or left of the center line. The exact number depends on the dimensions of the room and the geometry of the arrangement. This is how far the listener can move and still be able to restore the stereo effect by adjusting the balance.

9.10 Ambient noise

Another factor that affects the quality of sound is noise. In outdoor venues the noise problem has to be addressed early-on during the planning stage. Surrounding the venue by shrubs and trees creates a buffer that can reduce both incoming noise (usually from traffic) and outgoing noise from the venue to the neighboring area. Another way that noise can reach the venue is by refraction. This happens in areas prone to atmospheric inversion, as explained in section 2.8.1. This is another issue to be addressed in the planning stage.

Indoor acoustics are affected by a different set of noise problems that also have to be addressed early-on in the planning and design phases. For external noise, a design that includes panels with the correct noise reduction coefficient (NRC) is important. The NRC is a number that tells us the fraction of the noise absorbed. For example, NRC = 0.1 means that 10% of the incident sound is absorbed and 90% is reflected back; NRC = 0.8 means that 80% of the sound is absorbed and 20% is reflected. In addition to external noise, ventilation and air-conditioning systems can be major contributors of noise, both in large halls and small rooms. The noise is essentially guided into the space by the ductwork, especially metal ductwork. Some of the noise is generated by the ductwork, particularly if there are sharp bends, and also by the registers.

Similar noise issues exist in small music rooms and home recording studios, with additional problems from plumbing, resonance from window panes, doors and other. Some houses may have effective noise reduction paneling from the outside, but very little in terms of blocking indoor sources of noise. As most people realize when trying a home recording session, the amount of noise in the house that finds its way to the recording microphone is often amazing. Depending on one's specific goals, the situation can be remedied to some extent by adding soundproofing panels to the walls and the ceiling. The panels are more effective if they are not attached directly to the wall, but have a gap in between. Another important step is to isolate the noise from wooden floor boards and door gaps. The problem here is that whatever the remedy, the floor must be rigid enough to support people and equipment. In most cases the result is a raised floor, which may require additional alterations with doors and heating/air conditioning registers, and so on.

9.11 Further discussion

9.11.1 Arranging microphones for stereophonic recording

There are different ways of arranging the microphones for stereo recording. A common method is to have two cardioid microphones spaced 3 m (about 10 ft) apart (see section 7.4.2.) The pair must be located in front of the stage, and at a distance comparable to the lateral extent of the group performing on the stage. For instance, if the distance between drum and guitar is 4 m (12 ft) then the microphones should be about 4 m in front of the group. Another method uses two cardioid microphones on the stage, separated by a small distance, 30 cm or less (less than 1 ft), with their axes of maximum pickup (the 0° line in figure 9.6(b)) forming an angle of about 90–100°. One of the microphones should be pointing to the right half of the performing group and the other to the left.

9.11.2 Surround sound

A surround sound system adds three (5.1 system), four (6.1 system), or five (7.1 system) loudspeakers plus a subwoofer (the '.1' indicates subwoofer) to the L and R of the stereophonic system of figure 9.6(b). Depending on the number of additional speakers, one of the additional speakers can be located front-center, two on the sides, and two behind the listener, providing the 'surround' sensation. The signals to the additional speakers are a combination of the L and R sounds, or can be used exclusively for sound effects, if the system is used for video. The signal to the additional speakers is electronically delayed compared to the sound of L and R, to optimize the surround effect at the point of the listener.

9.11.3 Resonance in indoor spaces

In section 6.9 we found that the frequency modes for a pipe closed at both ends are

$$\text{frequency} = (n \times 343)/(2 \times \text{length of pipe})$$

where $n = 1,2,3,\ldots$ and so on. We can assume for simplicity that a closed room or hall has a cylindrical shape[3]. Let us assume that the closed pipe has length equal to the long dimension of the room, say 6.86 m (about 20 ft), and calculate the first few frequencies that can build up. Using the above equation, we find

$$25, 50, 75, 100, 125, 150 \text{ Hz, and so on.}$$

Note that the frequencies are spaced apart by 25 Hz, which as expected, tells us that the *spacing* of the overtones is equal to the fundamental frequency. In a large concert hall, say 50 m long (150 ft) the fundamental frequency is 3.4 Hz; so, the overtones are spaced apart by 3.4 Hz, which is much smaller than the spacing of the frequency modes in a small room. Of course, the hall has sidewalls, floor, and ceiling, which add more frequency modes. The result is that the spacing of the frequency modes in a large hall is so small that we essentially end up with a *continuum* of frequency modes. This means that there are no 'preferred' frequencies that will build up excessively in a large hall. This is not so for a small room.

Figure 9.7 shows the first five modes in a room 5 m × 6 m × 3 m (15 × 18 × 9 ft). We note that the spacing of the frequency modes remains significant, even when we include reflections from all sides of the room. Keep in mind that if the reflections can build up a standing wave of some frequency, then this frequency is also a resonance frequency, meaning that in small rooms some discrete frequencies will resonate, and the effect may be quite noticeable. Experimenting with the orientation of the

---

[3] Again, this is a model, just to help us get started. The results of using a more realistic 3-D model are shown in figure 9.7.

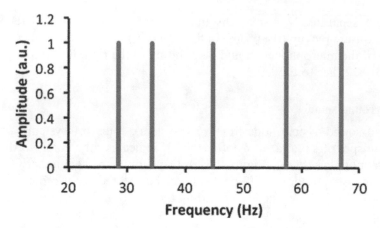

Figure 9.7. First five modes in a room 5 m × 6 m × 3 m. For simplicity we assume that all the modes have the same amplitude. Amplitude is in arbitrary units (a.u.).

speakers or the placement and orientation of furniture may be helpful in reducing the resonances in this situation.

From the discussion of closed pipes, we know that the walls represent antinodes for the pressure, meaning that the oscillation of the acoustic pressure is at maximum. This is true of small rooms as well. The air pressure oscillation is highest at the walls. Alternatively, if we want the loudspeakers to set the pressure oscillation in the room efficiently, placing them near the walls would be the best spot. This is a good reason for placing the speakers near the corners of the room.

9.11.4 Anechoic chambers

From our discussion it is clear that the characteristics of the environment affect the quality of the sound we produce or record. In many cases it is important to separate out the effect of the environment altogether. For example, to determine the pickup pattern of a microphone (see section 7.4.2) one has to measure the response of the microphone when the sound source is at *different directions to the axis of the microphone*. In doing so, the microphone responds not only to the direct sound, but picks up all the reflections of the sound from the walls in the room, which are coming from *all* directions. This could be a problem, especially if the room is small. For specialized work, one needs to eliminate these reflections, and of course minimize ambient noise.

Anechoic chambers (meaning chambers with no echo) use foam material such as that shown in figure 9.8 to cover the walls and ceiling. The surface of the foam has wedges, typically about 5 to 10 cm wide (2 to 4 inch) 5 to 10 cm tall in panels of various dimensions. The wedges can run the length or height of the wall. In other designs they are pyramid-like, or the inside of an egg carton. In any case, the basic function of these panels relies on making the incident sound wave reflect back and forth within the *valleys* of the foam, and the backing wall. With each reflection, some energy is lost, and the result is a significant reduction of the reflected sound.

Figure 9.8. Acoustic wall panels.

9.12 Equations

The Sabine equation

The reverberation time can be calculated from the Sabine equation:

reverberation time = 0.161 (volume of room/total absorption).

Each material has its characteristic absorption coefficient. This means that to find the total absorption by the surfaces in a room we need to consider all the surface materials separately. For example, we will consider an *empty* room with dimensions $5 \times 6 \times 3$ (width × length × height, all in meters). The volume is $5 \times 6 \times 3 = 90 \text{ m}^3$.

We will assume hardwood floor, drywall, and acoustic tile ceiling. The surface areas (with the typical absorption coefficients in parentheses) are:

- Floor $5 \times 6 = 30 \text{ m}^2$ (absorption coefficient = 0.3)
- Ceiling $5 \times 6 = 30 \text{ m}^2$ (absorption coefficient = 0.7)
- Two side walls of $5 \times 3 = 15$ plus two side walls of $6 \times 3 = 18$ or 66 m^2 total (absorption coefficient − 0.1)

To find the total absorption, we multiply the total surface for each material times the absorption coefficient of the material, and add. We find[4]

$$0.3 \times 30 + 0.7 \times 30 + 0.1 \times 66 = 36.6 \text{ m}^2 \text{ or \textbf{sabins}}.$$

Using the above values in the Sabine equation we find

$$\text{reverberation time} = 0.161 \times (90/36.6) = 0.4 \text{ s or } 400 \text{ ms}.$$

Note that in the above example we used meters to measure the dimensions of the room. The calculation is the same for measuring in feet, except that the coefficient 0.161 in the Sabine equation should be changed to 0.05 instead. Note also that the absorption coefficient values used are 'typical'. The absorption coefficient depends on the frequency as well, hence the dependence of the reverberation time on the frequency.

9.13 Problems and questions

1. A small lecture hall is 10 m long. Here we will ignore the reflections from the sidewalls, floor and ceiling.

---

[4] The unit **sabin** is commonly used, in honor of Wallace Sabine (1868–1919) who pioneered concert hall acoustics.

   (a) How long does it take for the direct sound to reach a student sitting in the middle of the hall?

   (b) How long does it take for the sound to complete 1 round trip between front and back walls?

   (c) Suppose the walls are covered with a material that absorbs 50% of the incident sound. What fraction of the sound energy is lost after 1 round trip?

   (d) Suppose that the walls are covered with a material that absorbs 10% of the incident sound. What fraction of the sound energy is lost after 1 round trip?

   (e) Which wall material would make the reverberation time of the hall longer?

2. A concert hall is 34 m long. Here we will ignore the reflections from the sidewalls, floor and ceiling.

   (a) How long does it take for the sound to complete 1 round trip between front and back walls?

   (b) Suppose that the walls absorb 10% of the incident sound. What fraction of the sound energy is lost after 1 round trip?

   (c) Other things being equal, will a longer hall have longer or shorter reverberation time?

   (d) Is the sound in this hall more likely to have 'intimacy' compared to the smaller hall of problem 1? Explain your answer.

3. Use the model of a pipe closed at both ends to compare two different rooms A and B. Room A has dimensions 5 m × 6 m × 3 m. Room B has dimensions 4.5 m × 4.5 m × 4.5 m.

   (a) Find the frequency of the first vibrational mode for each dimension of room A.

   (b) Find the frequency of the first vibrational mode for each dimension of room B.

   (c) Which of the two rooms has more resonances?

   (d) Which of the two rooms is better for listening to music? Why?

   (e) Does any of the frequencies found in part (a) agree with the values shown in figure 9.7?

4. Our voice sounds different in small enclosed spaces. That is why some people like to sing in the shower. Explain why, using a space of dimensions 1 m × 1 m × 3 m (close to the dimensions of an enclosed shower.)

IOP Publishing

The Physics of Sound Waves (Second Edition)
Music, instruments, and sound equipment
Panos Photinos

Chapter 10

Sound recording and reproduction

10.1 Introduction

Commercial recording and reproduction of sound dates back to the late 1800s when Thomas Edison patented his phonograph. The early models could both record and play back using rubber based or wax cylinders. The disk format was introduced in the early 1900s, and used for playback on gramophones. The gramophone disks were made of shellac originally and then from a polyvinyl chloride base, hence the term 'vinyl.' The early gramophones used spring driven motors, which required frequent cranking. By 1940 fully electrical units with electronic amplifiers replaced the older designs.

While records are still in production, and for some people vinyl remains the favorite medium, other recording and playback innovations have become popular as well. A very successful invention used magnetic media (magnetic wire, then magnetic tape). In the early 1980s, compact disks, or CDs were commercialized, and quickly became the preferred medium for about two decades. In this chapter we will describe the basic principles of some analog and digital recording and reproduction systems. Some of the concepts used here, like induction, frequency bandwidth, distortion, were introduced in chapter 7.

10.2 Gramophones

In a gramophone record the signal is recorded on a vinyl disk. The disks have grooves spiraling from the outer end towards the center. The walls of the grooves make an angle of about 45° to the vertical, and the sound signal is impressed as ridges on the walls, as shown in figure 10.1.

In a stereo system, there are two channels of signal, one on each wall of the groove. The sound signal to be recorded is converted into an electrical signal, which sets a **cutting stylus** into vibration. The vibrations of the stylus are engraved on a coat of shellac or lacquer. The grooved surface is then coated with a metal film, so the metal film becomes a 'negative' of the groove pattern. The metal negative is used as a print, to impress grooves on a vinyl record. For playback, a **stylus** is in contact with the moving groove pattern, and the ridges on the walls make the stylus move laterally (right–left) as well as vertically (up–down). In the early models, the stylus was a needle and the vibrations were fed to a diaphragm that produced audible sound with a horn (see opening figure of chapter 7) or without a horn (see also figure 10.5(b)). In some models the sound was heard using a stethoscope, which in a way makes the stethoscope a precursor of today's earphones.

The record is placed on a rotating **platter** (or **turntable**, although some use this term to refer to the whole gramophone unit). The platter is usually heavy (1–3 kg, or 2–6 lb) and the mass is distributed mostly at the rim, to produce high moment of inertia, like a gyroscope, which helps keep the rotation speed steady. The platter is

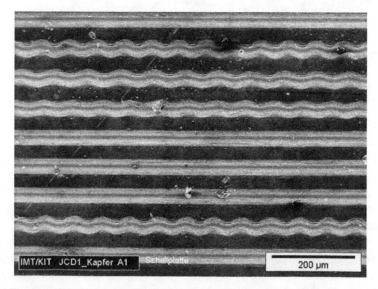

Figure 10.1. Microscope image of the groove in a gramophone record. Image credit: Klap (https://commons. wikimedia.org/wiki/File:Langspielplatte.jpg).

set to rotation by an electric motor. In one design, the motion of the motor's shaft is transmitted to the platter by a belt (**belt drive**). In another design, the shaft of the motor is directly attached to the axis of the platter (**direct drive**).

10.3 Magnetic tape deck

Recording sound signals on magnetic material was commercialized in the mid-1940s and became very popular with the introduction of the compact cassette in the early 1960s. The early modes used a magnetic wire, until the introduction of the plastic magnetic tape. The method is based on the interrelation between electric signals and magnetic fields, which was discussed in section 7.2.7b, and on the interaction between magnets. In a cassette the plastic tape is coated with finely powdered magnetic material such as iron oxides or chrome dioxide. The direction of magnetization of these tiny magnets can be aligned by the action of a magnetic field. Once aligned, the direction of magnetization of the particle remains unchanged, even after the aligning magnetic field is removed.

Figure 10.2 schematically illustrates the structure of a recording/playback **head**. The signal to be recorded is fed to a coil (red) wound on an iron ring (blue). The current in the coil produces a magnetic field (green lines) in the ring, which 'mimics' the incoming electric signal. A small gap in the ring filled with non-magnetic material (usually gold) 'kicks out' some of the magnetic field from the ring towards the tape. This field aligns the tiny magnets on the tape (indicated by the brown color band at the bottom of the figure), which is moving past the head at constant speed. The pattern of alignment of the magnetic powder follows the pattern of the input signal. In playback the process is basically reversed: as the tape is run past the head, the magnetic field from the powder in the tape produces a magnetic field in the ring,

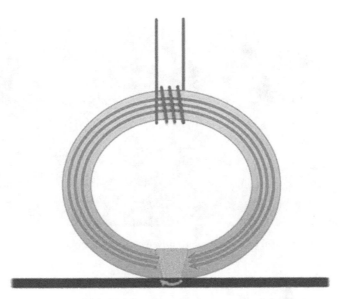

Figure 10.2. Structure of a recording/playback head of a tape player.

which in turn induces a voltage between the ends of the coil. This voltage is fed to the amplifier. Some units use the same head for recording and playback. Better units have separate heads for recording and playback. In these units the recording head has a wider gap (about 4 microns) than the playback head (1 micron)[1]. In addition, an erase head is used to make the alignment of the particles random before recording again. Random alignment means that the sum total of the magnetic field of the particles over a distance equal to the gap of the head is zero. The tape is moved past the head using a system of a motor-driven **capstan**, and a rubber wheel pressing against the capstan, shown in figure 10.3. The tape lies flat between the rubber wheel and capstan. It is important that the tape remains flat against the head and moves at constant speed. This is why some expensive units have two additional motors to keep the tape tight. For stereo sound, the head must have two tracks, one for each channel.

The magnetic pattern recorded on the tape is essentially a waveform representation of the sound signal (see section 3.3). If a graph of a waveform is more spread out horizontally, one can see the oscillations more clearly, in other words have more resolution. In the same way, if the tape moves fast, the pattern is more spread out, and the performance is improved, especially at high frequencies. Professional units use speeds of 38 cm s^{-1} (15 inch s^{-1}) or higher.

Tape decks suffer from noise (hissing), which comes mostly from the fact that the magnetic particles are finite in size. When the tape is unrecorded, or the contents of the tape have been erased, the magnetization of individual particles points in random directions. But since the particles have a finite size, the playback head reads each individual particle and the result is a weak random signal of high frequencies, which produces a characteristic 'hiss.' Most commonly, the decks come with a noise reduction system—for example, DBX or Dolby—to correct the noise problem.

Figure 10.3. A tape head (right) and motor driven capstan system and rubber wheel (left).

---

[1] 1 micron is one millionth of a meter.

10.4 Compact disk

The compact disk (CD) became commercially available in 1982, and its popularity increased rapidly, especially with the introduction of portable CD players in 1984. The first CDs were read-only. Recordable CDs were released in 1990, which increased their popularity even more. The CD essentially reflects light, and the amount of the reflected light can have two levels: high or low. As discussed in section 8.2, this two-state recording is essentially a binary signal, consisting of a sequence of 0 s and 1s corresponding to the ON and OFF levels of the reflected light. The binary signal is impressed on a clear polycarbonate surface, as a series of **pits** of uniform depth. The pitted surface is coated with a layer of reflecting material (usually aluminum), and another layer of protecting coating is applied on top of the aluminum. The sequence of layers and a pit pattern are illustrated in figure 10.4(a).

To read the signal, a laser beam (red in figure 10.4(b)) coming from the bottom of the polycarbonate is focused on the reflective aluminum layer. If the spot of the laser beam is over a flat part of the pattern, the amount of reflected light stays the same. This is the binary 0 state. But when the spot crosses an *edge* of a pit the amount of reflected light drops (see section 10.6 for more details). This is the binary 1 state. The reflected light is picked up by a system of light detectors (called **photodiodes**), and is interpreted and transformed to an audible analog signal, as discussed in section 8.3. The pits are impressed in an outwardly spiraling pattern, much like the inward spiraling pattern in vinyl records. A motor turns the CD, and a separate mechanical system keeps the laser and the photodiodes that read the reflections on the correct position below the pit pattern.

Note that in this scheme there is no physical contact between the device that reads the information and the surface that contains the information; in other words, the CD does not wear out with use, as is the case with records and tapes. Another advantage is that the signal is binary, which means that noise from other components of the system, like vibrations from the motor, does not affect the reading of the signal.

10.5 Compressed and uncompressed formats

Typically, a CD will store about 70 min of music. The code used by the manufacturer, the so-called CDDA, can be read by a CD player or a computer disk drive.

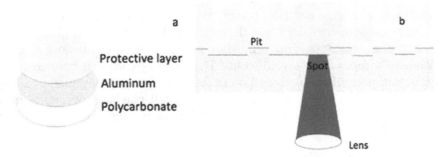

Figure 10.4. (a) CD layers. (b) Laser beam focuses on the pit pattern.

The contents of a CD can be 'ripped' by computer software. Microsoft converts the CD content into 'wav' format (indicated as '.wav' in the file name extension) and Apple uses the 'aif' format ('.aif' or '.aiff' file extension). In these formats the data are **uncompressed**, in other words, they include all the information as recorded, and because of that, they are very large. Recall from section 3.2 that we need 7.2 million voltage measurements, or samples as they are called in digital, to make 3 min of play. In digital audio systems, these samples are usually 16-bit words. Also, for a stereo CD we need two channels, right and left. Putting these all together, we need 7.2 million per 3 min × 2 channels × 16 bits = 230 million bits. In terms of bytes[2], we need 29 MB per 3 min of music, or roughly 10 MB min^{-1} of song, which applies both to the 'wav' and 'aif' formats.

Several schemes have been developed to reduce or **compress** the size of audio files. 'm4a' and 'MP3' are compressed formats. The 'm4a' format compresses the data but does not discard any information, so it is a **lossless** format. The popular 'MP3' format, takes up less memory space by using a sophisticated scheme that removes some of the sound frequencies. Recall from section 4.9 that if two frequencies are close enough, one frequency could mask the other. Therefore, some masked frequencies can be discarded without affecting the sound quality noticeably. 'MP3' is a **lossy** format (as opposed to the lossless format) because some of the original information has been omitted, and one cannot convert it back to the 'wav' or 'aif' original. In terms of memory space, the 'MP3' uses about 1 MB min^{-1}, which is ten times less than the uncompressed 'wav' and 'aif' formats.

10.6 Semiconductor storage devices

Advances in computer storage technologies led to the development of semiconductor storage devices, with storage capabilities far exceeding those of CDs. An added advantage is that semiconductor storage devices have no moving parts, meaning that there is no need for a motor. Also, the semiconductor storage devices are much smaller, lighter, and more durable. The transistor is the heart of these devices. Recall from section 8.3 that the transistor can act as a switch, so it can have two states (ON or OFF) that can be used for binary-logic devices. In the case of memory devices, millions of transistors[3] integrated into a small chip are used for storing huge amounts of data (billions of bytes). The storage is 'permanent' (as opposed to 'volatile'), meaning that the data remain in the memory without need of constant refreshing, which would require some electrical power, like a battery. Permanent does not imply that the data cannot be erased. In fact, the data stored in a memory card can be erased and re-written, and the data is not limited to audio, as the card can store all kinds of files, including video, pictures, and so on. The memory card can be used either as separate storage on a device (such as flash or thumb drive on a computer, or a chip on a tablet, and other) or it can be packed with a small playback device, as in MP3 players. In this case of course, a battery is required to run the player.

---

[2] Recall from section 8.1, that 1 byte is 8 bit.
[3] The metal-oxide semiconductor field emission transistor (MOSFET) is the most common at this time.

In summary, semiconductor memory storage has all the advantages offered by the CD with the additional advantages of more memory and smaller size. As a result, semiconductor memory has gained great popularity as *the* audio storage medium, while the other media (records, tapes, and CDs) have declined.

10.7 Further discussion

10.7.1 Record player cartridge

In modern units, the stylus is attached to the **cartridge**, shown in figure 10.5(a), which in most cases is similar in principle to the dynamic microphone. In one design, the stylus' far-end (opposite the one contacting the record's surface) has a small coil attached to it. The coil moves in a magnetic field which is on the cartridge, and so it produces an electric signal at the end of the coil's wires. This design is the **moving coil** (MC) cartridge. In another design, the magnet is attached to the far-end of the stylus, and the magnet moves inside a coil which is on the cartridge. This is the **moving magnet** (MM) cartridge. The cartridge is attached to the **tone arm**, which has a delicate balancing mechanism that adjusts the force exerted by the stylus on the surface of the record, typically less than 5 g (0.18 oz). Figure 10.5(b) shows the cartridge of an older unit. In this design, the stylus is a needle, and the aluminum foil visible through the holes is actually the 'loudspeaker' of the unit.

a   b

Figure 10.5. (a) Gemini cartridge on a TEAC turntable. (b) Cartridge of 1930s 'His Master's Voice' Suitcase Phonograph. The needle is at bottom center.

10.7.2 Reading data from a CD

The depth of the pits in a CD is selected to be equal to *one quarter* of the wavelength (λ) of light in the polycarbonate material of the CD. Figure 10.6(a) shows the light spot at three different instances. When the laser beam hits the edge of a pit, part of the light reflects from the bottom of the pit, and part of the light reflects from the top of the pit. In other words, part of the light has traveled an extra distance of $\lambda/2$. Recall that one half of a wavelength means that the two parts of the reflected beam are 180° out of phase. The phase difference causes some light cancellation as the laser spot crosses an edge. The pink arrow in figure 10.6(b) indicates this reflection. Cancellation does not occur when the spot hits the flat parts at the top or bottom of the pits, as indicated by the red arrows in figure 10.6(b).

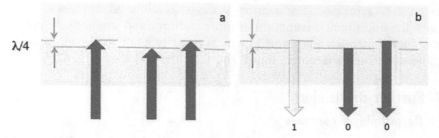

Figure 10.6. (a) Laser spot incident on a CD at three different instances. (b) Light reflected from the aluminum layer. The weak reflection (pink) corresponds to binary 1 and the strong reflections (red) correspond to binary 0.

10.7.3 Mastering and mixing

In the early days of sound recording, the phonograph horn (see opening image of chapter 7) was used both as a loudspeaker for playback and as a microphone for recording. All the sounds from instruments and voices were recorded at once. There was no room for editing, or cut and paste, so, the entire piece from beginning to end had to be recorded in one take. Of course, this meant a lot of rehearsing and several takes, if that was an option. The most acceptable take was then the **master** copy, from which the records were produced. The situation changed considerably over the years. The major advancement came with the advent of magnetic tape recording, which besides cut-and-paste, allowed several significant improvements. For example, different instruments and voices could be recorded in separate *tracks*, at different times, even different continents! After recording the tracks, the next step is to combine the tracks to make the final master. This is a task that requires a combination of engineering skill and artistry. The basic process involves **mixing** the tracks using a sound mixer, as discussed in section 7.8.7. The challenge however is how much volume of each track to add to the mix. For example, how loud should the bass guitar be compared to the vocals? Also, for each instrument/voice, the engineer may decide to use an equalizer (see section 7.5) to change the frequency proportions in the mix, and make the sound warmer or livelier, and so on. Another significant step is to make sure that the sounds are mixed in a way that ensures a wide dynamic range. Which means to make sure that the loud and soft sounds are all heard without distortion.

10.7.4 Re-mastering

Re-mastering refers to improving an existing recording, or the master used for a given recording, if that is available. This has been done extensively to old gramophone recordings of various genres. The analog sound is imported into a digital audio workstation (DAW), which can be a stand-alone unit, or computer software[4]. If the 'master' is just one track, then re-mastering is essentially limited to

---

[4] Such as Ableton Live, GarageBand, and other. There are also several free software and apps.

noise and hiss removal, some editing of the frequencies, and the overall volume, which makes the sound 'come alive.' Actually, this 'coming alive' means mainly loud recording with improved signal-to-noise ratio, but very little in terms of increase in dynamic range. If the 'master' is recorded on tape or in digital form, and there is more than one track, then one can do more manipulations, depending on the number and content of the tracks. There are of course some limitations to what can be done. For example, if the original is in lossy format, like MP3, then restoring the original content is not doable (see section 10.5). In re-mastering, as in mixing, one of the objectives is to increase the dynamic range without **distortion**.

10.7.5 Distortion

One way to distort a sound is to add frequencies that are not part of the original sound. These extraneous frequencies can be created if one of the components used (microphone, amplifier, recording device, amplifiers or loudspeakers) is driven beyond its linear range, as shown in figure 7.7. To illustrate the point, suppose that we use a magnetic tape for recording. There is a signal level where the magnetization of all the particles gets to perfect alignment. If the signal goes above this level, the magnetization will not align any better, because it is already aligned as much as possible. So, the response reaches the maximum and flattens out. This 'flat' part, brings in frequencies that were not in the original input signal, and that causes the distortion of the sound.

The so-called *clipping* (section 7.8.6) is an extreme case of distortion. The usual culprit is the gain level. For example, if the gain is set high to capture the softer sounds, during recording or re-mastering, then the louder sounds may become distorted. Audios 10.1 and 10.2 capture a pure tone of 220 Hz recorded at medium gain and at very high gain. Figure 10.7 shows the corresponding amplitude spectra. Note that in figure 10.7(b) there is a number of added frequencies, including the harmonics of 220 Hz, and other frequencies that are unrelated to the original pure tone of figure 10.7(a).

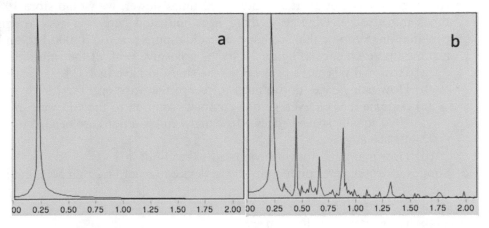

Figure 10.7. (a) Amplitude spectrum of pure 220 Hz recorder at medium gain. (b) Same tone recorded at high gain. Frequency in kHz. Vertical axes in arbitrary units.

Audio 10.1. Pure tone of frequency 220 Hz recorded at medium gain. Available at https://iopscience.iop.org/book/978-0-7503-3539-3.

Audio 10.2. Pure tone of frequency 220 Hz recorded at high gain. Available at https://iopscience.iop.org/book/978-0-7503-3539-3.

10.8 Problems and questions

1. List two advantages of magnetic tape recording over vinyl disk recording.
2. List two advantages of CD over magnetic tape recording.
3. List two advantages of semiconductor storage devices over CDs.
4. We could increase the storage capacity of CD (and vinyl disks or tapes) by rotating the CD slower. If we rotate 100 times slower, we would store 100 times more songs! Obviously, there is a limit and here is a very rough explanation. Suppose that we have sinusoidal pure tone of 10 000 Hz, and assume that each edge in figure 10.4(b) is a positive peak of the sinusoid.
   (a) How many positive peaks does the sinusoid have in 1 s?
   (b) How many edges do we need to record the waveform for 1 s?
   (c) Assume that on average one complete rotation of the CD we can fit 5000 equally spaced edges. How many full rotations are needed to fit 10 000 edges?
   (d) How many full turns do we need to read-out in 1 s?
5. Suppose we use a semiconductor storage device to record a 2 min MP3 audio file.
   (a) How many bytes do we need?
   (b) How many bits do we need?
   (c) Assume that the dimensions of our storage device are 1 cm × 1 cm. How far apart should the bits be on the semiconductor chip?

IOP Publishing

The Physics of Sound Waves (Second Edition)
Music, instruments, and sound equipment
Panos Photinos

Chapter 11

Percussion, wind and string instruments

Image credit: Janet Bocast.

11.1 Introduction

The use of musical instruments dates back to the Palaeolithic age. Archaeologists have found flutes constructed about 40 000 years ago from animal bones. Other instruments made from materials that disintegrate, such as drums made of animal skin, may have been in use at that time as well, but are very difficult to identify in finds. Many musical instruments evolved from early forms that date back to ancient China, Mesopotamia, Egypt and others. Some instruments, such as the piano were invented in the past few centuries, and have no ancestors in antiquity. The development of musical instruments, the shape, and the selection of the materials, was based on intuitive understanding of the properties of sound, and the process of trial and error. The famous Stradivarius violins are a case in point. These violins were made between approximately 1666 and 1737 AD by the Stradivari family in Cremona, Italy, and in the opinion of many experts, they remain unsurpassed in terms of quality. Several explanations have been proposed attributing the Stradivarius quality to a 'secret' in the wood, or the glue or the varnish and so on.

In chapters 3 and 6, we learned that a musical tone played on a given instrument has a characteristic spectrum of frequencies, and that the spectrum depends on many factors, including the shape of the instrument, the materials the instrument is made of, and the way the note is played. For example, a steel string sounds different than a nylon string, and a string plucked in the middle sounds different than when plucked closer to the fixed end. Spectral analysis of the sound of musical instruments, and knowledge of the properties of materials used to make musical instruments are essential in identifying the characteristics that make a good quality instrument.

Musical instruments are categorized in five main categories, which are **idiophones**, **membranophones**, **chordophones**, **aerophones**, and **electrophones**. Idiophones include instruments such as chimes, cymbals, marimbas, maracas, and others. What these instruments share in common is that the sound is produced by striking, rubbing, or shaking the instrument. Membranophones are drums of various kinds. Chordophones are instruments that use strings to produce sound. Aerophones are instruments that use air, such as flutes, clarinets and the like. Electrophones produce sound using electricity, like synthesizers. This is essentially the Hornbostel–Sachs classification system. In this chapter we use concepts developed in chapters 3 and 6 to understand the function and characteristics of the three basic sections of a symphony orchestra: the *string*, *wind* and *percussion* instruments. In the next chapter we will discuss the human voice and keyboard instruments.

11.2 Basic functions of musical instruments

The heart of a musical instrument is a vibrating system, the so-called **vibrator**. The vibrator can be a string, an air column, a metal rod, and so on. The key here is that the vibrator can sustain standing waves. The player supplies the energy to the vibrator, and part of this energy is radiated to the surrounding air in the form of audible sound. In most instruments the player has some control on the basic characteristics of the sound produced by the instrument, which includes pitch, loudness, duration, and timbre. Depending on the type of vibrator, we have the following major categories of instruments:

wind instruments (where the vibrator is an air column, as in a flute);

string instruments (such as the violin, use the vibrations of strings);

percussion instruments (the vibrator is a rod, as in chimes, or a surface like the membrane of a drumhead).

The process of starting the standing waves in the vibrator is called the **excitation** mechanism. So, one can set a string to vibration by plucking it or running a bow across it. Depending on the type of excitation, and other characteristics, one can distinguish subcategories of instruments. As will be discussed below, a wind instrument can be classified as a *reed* instrument (for example, the clarinet) or *brass* instrument (for instance, the trumpet) and so on. In a similar way, the string instruments can be *plucked* or *bowed*. In the following sections we will describe selected examples from each of the three major categories.

11.3 Percussion instruments

It is believed that percussion instruments are the oldest, which is reasonable in view of the simplicity that characterizes many percussion instruments. Percussion instruments comprise a wide range of rather dissimilar instruments, their unifying feature being that the sound is produced by striking the vibrator by hand or mallets, or other means. Even within a subcategory, drums for instance, the variations are too numerous to mention. Some percussion instruments use rods or tubes as their vibrator, others use stretched membranes, metal plates, and other.

11.3.1 Vibrating rods and tubes

The vibrational modes of rods and tubes were described in section 6.10. Recall that the fundamental produced by a vibrating rod, say an aluminum rod, depends on the length and the thickness of the rod. The frequency of the fundamental mode is lower for a longer rod, a lot like what happens with strings. In the case of rods, however, the relation between the fundamental frequency and the length of the rod is slightly different. To make the tone one octave higher we need to make the rod four times shorter. Another difference is the relation between frequency and thickness. A thin rod bends more easily than a thick rod, therefore the thin rod will have lower fundamental frequency than a thick rod of the same material and length. This behavior is opposite to what we found for strings. One common characteristic of rods and tubes is that the overtones are not harmonic, that is the overtones are not integer multiples of the fundamental frequency.

Chimes or **tubular bells** are essentially a set of 10–20 brass tubes of various lengths, hanging vertically suspended from the top. The top has a protruding rim, which is the striking point. A plastic hammer is commonly used for striking the tubes. The sound we hear is actually the combination of the fourth, fifth and sixth modes, which happen to have frequencies in the ratio 2:3:4. As a result, we hear a missing partial or virtual pitch[1] which is one octave lower than the fourth mode. As the tubes are suspended essentially freely, they can oscillate for a long time, therefore a foot pedal is used to damp out the oscillation.

The **marimba** is a popular instrument that originated in southern Africa. The vibrators of a marimba are horizontal wooden bars suspended near the two ends by strings. The bars are struck with mallets to produce a fundamental frequency, and non-harmonic overtones. Under each bar there is a vertical pipe closed at one end. The length of the pipe is selected so that the pipe resonates at the fundamental frequency of the bar. The rods of the modern marimba are arranged following the pattern of the piano keyboard and the tonal range of a marimba is usually over three octaves.

Recall that for rods the frequency dependence on the length is stronger than in strings. So, if we want to go one octave lower, the length must increase by a factor of four. If we start with a bar of 0.2 m (about 8 in) for the highest note, then the note one octave lower would have a length of 0.8 m (32 in) and the next octave would require a bar 3.2 m (over 10 ft) long, which would be unpractical. For this reason, an arch is carved out of the bottom side of the bars. In this way the bars are thinner at the middle, and the overall stiffness is reduced, leading to lower frequencies of oscillation. The **xylophone** and the more complex **vibraphone** operate essentially on the same principle.

11.3.2 Vibrating plates

Another important group of percussion instruments uses a two-dimensional metal surface as the vibrator. Some of the general trends that are observed in the mode frequencies of strings and rods apply to surfaces as well. For example, the larger surfaces produce lower frequencies; a stiffer metal plate produces higher frequency. The mathematical description of vibrating surfaces is more complicated than it is for strings and air columns. We will first discuss the vibration of metal surfaces with free ends. In doing so we will use the so-called Chladni patterns, which will help us understand the vibration of surfaces[2]. The patterns are created by sprinkling small particles, like sand, on a vibrating surface. The particles eventually settle at points of the surface that are not vibrating. In other words, the particles settle at the nodes, which in the case of a two-dimensional surface are *nodal lines*. Figure 11.1 shows

Figure 11.1. Chladni patterns in circular disk. (a) Vibration frequency is 2434 Hz. (b) Vibration frequency is 5670 Hz. (Attribution: Pieter Kuiper, Public domain, via Wikimedia Commons.)

---

[1] See section 4.9.
[2] After E Chladni, 1756–1827.

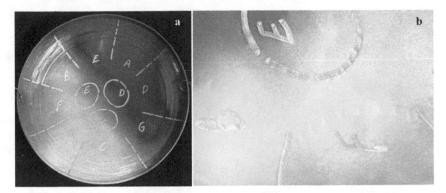

Figure 11.2. (a) Steel drum head. (b) Indentations defining note areas.

some of the patterns, that is, the vibrational modes, that can be excited in a circular plate, driven at the center. The frequency of the oscillation is indicated at the bottom right. In these patterns, sections separated by nodal lines move in opposite directions (see also discussion of figure 11.3). Note that at the lower frequency the nodal lines are circles. At the higher frequency, we have circles and straight radial lines crossing the circles. The patterns described can be sustained by circular plates with free ends, and are a good approximation of what happens in cymbals, gongs, and the like.

The **steel drum**, or **steel pan**, is essentially a concave metal surface. In the original version, developed in Trinidad, the metal surfaces were cut off 55-gallon oil drums, hence the name.

The surface has bumps each corresponding to a note, as shown in figure 11.2(a). Each bump is surrounded by an array of indentations, shown in figure 11.2(b) which define the vibration area for that note. The notes are tuned by the size and height of each bump, and are sounded by striking the note area with soft tipped sticks. The tonal range of a steel drum depends on the size of the pan, and for a large drum it is about two octaves wide.

11.3.3 Drums

The notion of nodal lines can also help us understand the vibration of drum surfaces. The situation here is slightly different from vibrating metal plates in that the rim of the vibrating membrane is clamped. In other words, the rim is a nodal circle. Figure 11.3 shows the first few modes for a flexible thin membrane. The colors indicate the direction of motion. For example, we can use blue to indicate the upward displacement of the membrane, and red to indicate the downward displacement.

As in the case of the vibrating metal plate, we have nodal circles centered at the center of the membrane. We also have straight nodal lines *through* the center of the membrane[3]. A pair of numbers labels the modes. The first indicates the number of straight nodal lines; the second number indicates the number of nodal circles. For example, the (0,1) mode has zero nodal lines through the center, and one nodal circle

---

[3] In figure 11.1, the disk is driven at the center, so the nodal lines do not go though the center.

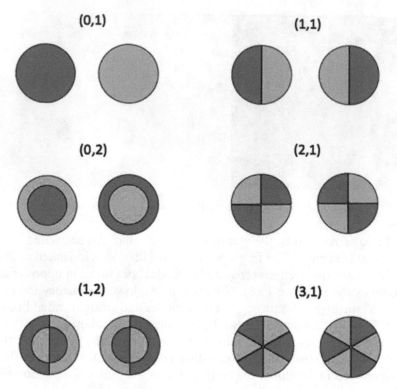

Figure 11.3. First few modes for a flexible thin membrane.

(at the rim). If the membrane moves downward (red) for the first half of a vibration cycle, then it moves upward (blue) in the second half of that cycle, and so on. In the (1,1) mode, we have one nodal line through the center and one nodal circle (the rim). If during the first half of a cycle, the left half of the membrane moves down and the right half moves up, then in the second half of the cycle the motion is reversed. The combined motion in this case, would be similar to Video 6.2(a). In the (0,2) mode we have no lines through the center and two nodal circles. If the inner part of the membrane is moving up, the outer part is moving down, and the motion pattern alternates every half cycle of the vibration. In the (3,1) mode we have three nodal lines through the center and one circle (the rim). Note that the nodal lines divide the membrane into six sectors, and that the direction of motion alternates between adjacent sectors. Looking at the other mode patterns, it is clear that the displacement of the membrane changes direction (up-down) as we cross any of the nodal lines (straight lines, or circles).

The frequencies corresponding to the modes shown in figure 11.3 are listed in table 11.1. Here we use the frequency of the (0,1) mode as our unit. The exact frequency of this mode will of course depend on the size and tension of the membrane. But suppose for example that it is 200 Hz. Then the frequency of the (1,1) mode is $1.59 \times 200 = 318$ Hz. In other words, table 11.1 lists frequency ratios.

Table 11.1. Frequency ratios corresponding to the vibrational modes shown in figure 11.3.

Mode	(0,1)	(1,1)	(2,1)	(0,2)	(3,1)	(1,2)
Frequency (Hz)	1.00	1.59	2.14	2.30	2.65	2.92

We see that the frequencies of the modes are not integer multiples of a fundamental, so the overtones are non-harmonic.

As we move to higher mode numbers, we have more nodal lines; in other words, the surface of the membrane is divided into more sections. Being smaller, these sections can move faster, hence higher mode numbers will generally correspond to higher frequencies, although the exact dependence of the frequency on the mode numbers is not straightforward. The frequency ratios listed in table 11.1 apply for a totally flexible thin membrane with no significant interaction with the surrounding air. Actual drumheads have some stiffness, and some thickness is necessary in order to provide strength. As a result, in an actual drumhead the frequencies of the higher overtones are larger than the values in table 11.1. Also, the interaction with the surrounding air slows down the vibration, especially for the lower modes; hence the frequencies of the lower modes are lower than the values in table 11.1. As a result, the frequency of the (1,1), (2,1), (3.1), and (4,1) modes come very close to being a harmonic series, and the sound of the drum has a discernible pitch.

As is the case of string instruments, the quality of the tone depends on how and where the membrane is struck. Looking at the mode patterns, we see that all the modes having nodal lines through the center have a node at the center; this means that if the membrane is struck at the center, the modes having circular nodal lines will be emphasized. On the other hand, striking near the rim emphasizes the modes having nodal lines through the center.

As the membrane is moving up, for example, in the (0,1) mode, it creates a pressure high. At the same time, the backside of the membrane is creating a pressure low. So, the sound waves created from the front and the back of the membrane are completely out of step. Because of diffraction, the sound bends, and eventually the sound from the back bends around and meets the sound from the front. As the two waves are 180° out of phase, significant cancellation takes place (see section 2.8.2). The result is a weak sound. To avoid cancellation, drumheads are mounted on **baffles** that prevent some of the sound from the backside reaching the front, or at least make that sound wave take a longer path to reach the front. In this case, the phase difference between the two waves is less than 180° and the cancellation becomes less significant. The drumheads need to be stretched uniformly, and the tension must be such as to produce the right pitch. This is usually done using five or more adjustable tension points to achieve proper tuning. The tension adjusting, or tuning rods, of a conga drum are shown in figure 11.4(a). Drumheads use animal skins or synthetic materials. In some cases the drumhead is covered with a layer of flexible paste (like bread dough) to shift the frequencies of the vibrational modes, or use skin with fur. An example is the tambourine shown in figure 11.4(b).

Figure 11.4. (a) Conga head with tension rods. (b) Tambourine from goat fur.

Figure 11.5. A typical drum set.

A typical modern drum set, shown in figure 11.5, usually includes a **bass drum**, a **snare drum**, and three **tom-toms**. The bass and the snare drum have two drumheads one on each end of their cylindrical body, the 'shell.' The tom-toms can have one or two heads. A set of gut or metal wires is stretched across the lower drumhead, giving the snare drum a rattling sound. Note also that the drum set includes a number of brass conical plates, the **cymbals**, with a dome at the center. The vibrations of a cymbal plate are similar to those of a metal plate. The basic cymbals are the **hi-hat**, the **ride** cymbal, and the **crash** cymbal. The hi-hat consists of two plates mounted on a stand, and is operated by pedal. The ride cymbal is a single plate mounted on a stand. It is struck by a drumstick, and has more sustained sound. The crash cymbal is smaller in size than the ride cymbal. It is struck, usually once and hard, as opposed to the ride cymbal that is struck rhythmically.

In western culture, the drums, and percussion instruments in general, have more of a supporting role. In some traditions, the percussion instruments or just drums alone, have the leading role, if not the only role, as in figure 11.6(a). This is common in many African cultures, the drummers groups from Burundi being the best known. All-percussion bands are very common in Polynesian cultures. The Fijian Lali, shown in figure 11.6(b) and its variations, is a carved log, struck by wood sticks.

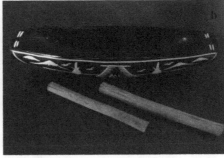

Figure 11.6. (a) A Burundian drummer band. (Attribution: Andreas31, CC BY-SA 3.0, via Wikimedia Commons.) (b) A Fijian Lali with unequal size drum sticks.

A band may consist of a dozed of various sizes of Lalis, to cover a wide tonal range, in the same way that steelpan bands (section 11.3.2) commonly include about a dozen of different size instruments, that together cover a tonal range of about 6 octaves.

11.4 Wind instruments

Wind instruments are essentially based on forming standing waves in the air column inside the **bore** of the instrument. Wind instruments have much more complicated shapes than the straight cylindrical pipes discussed in chapter 6. For example, the bore of some wind instruments like the flute, is cylindrical, but has holes on the side of the pipe. Others have a conical bore, like the clarinet. Others, like the trumpet, have a cylindrical bore that loops around, and transitions into a flared bell-shaped open end. As a result of the intricate shape of the air column, a wide range of fundamental tones and harmonics are allowed, which make these instruments richer. Although the formulas of sections 6.6 and 6.7 may not strictly apply to real wind instruments, pipes are useful as a model to understand wind instruments at the fundamental level. For example, how one can play different notes from a wind instrument, and what kind of tone range we can expect from a given instrument.

It is important to recall that the mode frequencies of a pipe (or a string for that matter) are also the resonance frequencies of the pipe (or string). If we manage to introduce into the pipe pressure waves of a wide range of frequencies, only the frequencies that match the mode frequencies of the pipe will build up standing waves, and only these frequencies will resonate, and eventually produce a tone. In other words, the pipe is *a filter of frequencies*. In the same way, exciting a standing wave in a wind instrument can be thought of as continuously injecting bursts of air, which is essentially noise. Noise contains all frequencies, as discussed in section 5.10. Of all the frequencies injected, the only ones that survive and resonate will be the ones that match the frequency modes of the instrument. A simple example is what happens when we blow sideways across the mouth of a bottle, discussed in section 6.14.1. Actual wind instruments use more intricate excitation mechanisms that allow more control over the characteristics of the sound produced. There are three main excitation mechanisms used in connection with wind instruments, which lead to three main categories, which are discussed next.

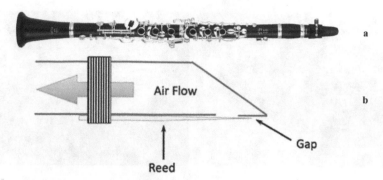

Figure 11.7. (a) Clarinet. Attribution: Yamaha Corporation, CC BY-SA 4.0, via Wikimedia Commons. (b) Schematic of mouthpiece with reed.

11.4.1 Reed instruments

One way to excite vibrations in a pipe is to blow onto a flexible **reed.** The air flow can set the reed into vibration, like what happens when a flag is waving. The reed is firmly attached to one end of the pipe, the **mouthpiece**, leaving a small gap that allows air to flow into the instrument. The structure of the mouthpiece is shown schematically in figure 11.7(b).

The player's lips seal around the mouthpiece (and reed) near the right end of the figure, blowing a continuous airstream. As the airstream flows through the narrow gap, the air begins to move faster, and the higher speed creates an upward suction on the reed[4]. If the reed is flexible enough, it will bend upward and seal the gap, and stop the airflow into the bore. The pressure at the gap at this point is low, and the gap will remain closed until some high pressure builds on the upper side of the reed and forces it open again. In other words, the mouthpiece end acts as a closed end of a pipe. The puff of air that entered through the gap consists of waves of many frequencies, which is essentially noise. When the waves reach the open end of the bore, as discussed in section 2.4.3 they will reflect back into the air column and travel towards the gap. After a few round-trips, the only frequencies that will survive are those that correspond to a pressure node at the open end and a pressure antinode at the gap. When the antinode at the gap reaches a pressure 'high' the pressure will push the reed down and open the gap. Once the gap is open, the airstream flows through the gap again starting a new cycle, and eventually a stable tone is established.

It is important to note that the 'open–close' cycle of the 'valve' action of the reed is controlled by the standing wave created in the pipe; in other words, the reed follows the resonant frequencies of the air column. This is an example of **feedback**, where the excitation (valve action of reed) is fed into a resonator (air column) and the wave selected by the resonator is sent back to the reed and in a way controls the timing of the excitation (valve action of the reed). In our example, the standing waves have maximum pressure variation (antinode) at the gap, and we have a pipe

---

[4] The creation of a pressure low by a fast moving air stream is known as the Bernoulli effect.

closed at one end (see section 6.7). So, wind instruments that use reeds act as pipes closed at one end, and include the **clarinet** and **saxophone**. Some instruments use double reeds, based on the same principle described above. Double reed instruments include the **bassoon** and the **oboe**.

11.4.2 Brass instruments

In brass instruments, like the **trombone**, the **trumpet**, and the **tuba**, the lips that are pressed closed against the cupped end of the mouthpiece do the 'valve' action. As the player blows, the sealed lips temporarily open and a puff of air is fed into the bore. This puff consists of waves of many frequencies, which are reflected back at the open end. After a few round-trips, the frequencies that correspond to a pressure node at the open end and an antinode at the mouthpiece will establish a standing wave. When the pressure variation at the mouthpiece reaches a 'high' it pushes the lips open, another puff comes in the air column, and the cycle repeats. The result is that the rate at which the lips open and close is controlled by the frequency of the standing waves in the air column, which means that a feedback loop is established as in the reed vibrations discussed above. The mouthpiece is closed most of the time, opening slightly and briefly as a puff is blown into the bore. Therefore, brass instruments are more like pipes open at one end (the bell) and closed at the other end (the mouthpiece).

11.4.3 Air-reeds

A third way to achieve the 'valve' effect is by blowing against a rigid sharp edge, as shown schematically in figure 11.8(a). Upon meeting the sharp edge, the airflow from the mouth can go to the right or the left of the non-vibrating edge in an irregular manner. When the airstream flows to the left of the edge, it enters the pipe, and as in the other two excitation mechanisms discussed above, some frequencies

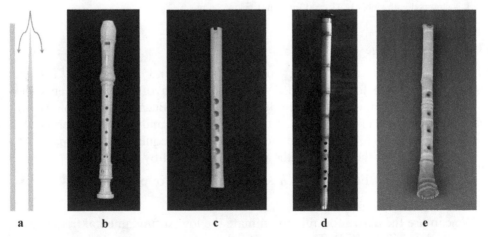

Figure 11.8. (a) Schematic illustration of the air-reed. The air stream, indicated by blue arrows, oscillates in and out of the bore. (b) Recorder. (c) Andean *quena*. (d) Turkish ney. (Attribution Alessandro dAgostini, CC BY-SA 4.0, via Wikimedia Commons.) and (e) Japanese *shyaku hachi.*

will build up a standing wave after a few round trips in the pipe. As before, the feedback will force the incoming air stream to oscillate to the right or the left of the sharp edge at a regular frequency imposed by the standing wave that builds up in the air column. This type of excitation uses the so-called **air-reed**, although there is no reed at all in this case. This excitation mechanism applies to many flutes and recorders. Figure 11.8(b) shows a recorder, (c) a *quena*, traditional flute of the Andes, (d) a Turkish *ney*, which is also popular in Egypt and elsewhere in the middle east, and (e) a Japanese *shyaku hachi*.

Figure 11.9(a) shows the schematic of a Native American flute. Here the air-stream is defined by a block that is strapped to the instrument. The air-reed or 'spliting-edge' is visible at the center of figure 11.9(b), just below the dark brown block.

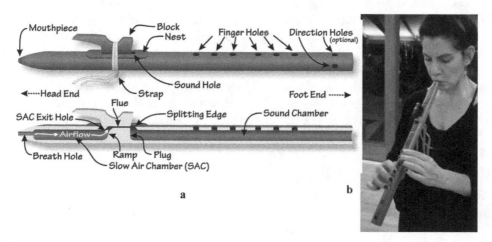

Figure 11.9. (a) Schematic of a Native American flute. (Attribution: ClintGoss, CC BY-SA 3.0 https://creativecommons.org/licenses/by-sa/3.0, via Wikimedia Commons.) (b) Hole at the bottom is not played. Block is strapped with ribbons. Air-reed forms just below the block.

11.4.4 Playing range

Figure 11.10 shows the playing ranges of some wind instruments. The piano keyboard is shown for reference. We note that each of the instruments shown covers three-plus octaves in different frequency ranges. Because the piccolo is shorter than the flute, it has the highest pitch. This can be understood in terms of our analysis of section 6.6, where we found that the mode frequencies in an open pipe are inversely related to the length of the pipe, as expressed by the formula

$$\text{Frequency} = n \times 343/(2 \times \text{length of pipe})$$

where $n = 1, 2, 3$, and so on.

We can use the above formula to estimate the lowest fundamental frequency of a flute, which is 262 Hz (C4). The length of a flute is about 0.66 m. Using the formula, for the fundamental ($n = 1$) we find the frequency 260 Hz, which is in good agreement with the actual frequency of 262 Hz.

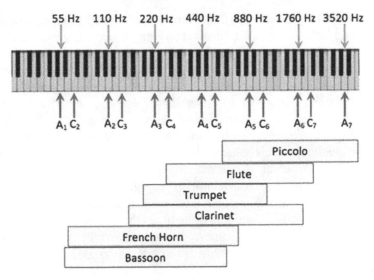

Figure 11.10. Playing ranges of some wind instruments.

Length of Air Column

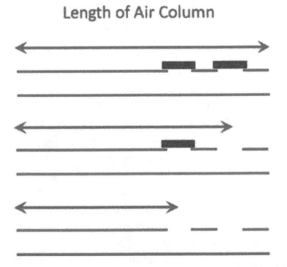

Figure 11.11. Effective length of the air column gets shorter as the holes on the side of the pipe open. The exact value of the length of the air column depends on the size of the tone hole.

The next, and probably most important, question is how do we get the range of notes shown in figure 11.10. According to our model, we need to change the length of the pipe. One way to do this is to open *tone holes* at the side of the pipe. Having an open hole on the side of a pipe in essence changes the 'effective' length of the vibrating air column, as shown in figure 11.11. When the holes are plugged (top) the length of the oscillating air column is essentially the length of the pipe. When the first hole on the right is open, the opening of the hole becomes an open end for the pipe, and a lot of the sound escapes from this hole. In other words, the length of the oscillating column or **acoustic length** becomes shorter, and the frequency increases. When the second hole is also open, the acoustic length becomes even shorter, and the frequency higher. Using properly spaced and properly shaped holes at the side of the pipe allows instruments such as the flute and the clarinet to play about three octaves. This tonal range covers about three-dozen tones, which is much more than one can count on their fingers. One way to cover this wide range is to take advantage of harmonics. By blowing hard, or **overblowing**, one can excite the second harmonic, which has twice the frequency of the fundamental, that is, one octave higher. This works well for the flute, which behaves like an open pipe, where by overblowing one can go from C4 to C5, which is one octave higher.

For the clarinet, the situation is not as straightforward, because the clarinet behaves like a pipe closed at one end, discussed in section 6.7. In these pipes the harmonic next to the fundamental has three times the frequency of the fundamental. In other words, by overblowing a tone in the clarinet we get more than one octave higher. So overblowing E3 for instance will get us to B4, which is one octave followed by a fifth[5]. Besides overblowing, mechanical keys, fingering, and blowing techniques are used to extend the range to over three octaves and produce all the tones in that range. The specifics are beyond the scope of this book, and the interested reader should refer to more specialized literature.

Much of the above discussion applies to brass instruments, with three main differences. First, instead of using holes to change the acoustic length, brass instruments use valves (like in a trumpet) or sliding pipes (like in a trombone). The principle is schematically illustrated in figure 11.12. Figure 11.12(c) shows an over-simplified setup that can change the effective length of the air column. When the valve is up (top of figure 11.12(c)) the oscillating air column builds up along the straight section of the pipe. If the valve is down (bottom of figure 11.12(c)) the air column follows the alternative curved path, which is longer. Figure 11.12(d) shows the method used in the trombone. The U-shaped pipe is inside the two straight sections, and can slide back and forth, changing the acoustic length accordingly.

The second difference is that in brass instruments the open end flares into a large bell, which is a significant difference from the straight pipe. As mentioned earlier, brass instruments behave like pipes closed at one end, and one would expect odd harmonics to form (like the clarinet). But because of the bell ending, the mode frequencies change, and in terms of overtones the instrument behaves more like an open pipe; in other words, it can have both odd and even harmonics.

The third difference is that the bell allows a smoother transition of the sound energy from the air column to the surrounding air. The result is that more energy can flow from the instrument, making it sound louder than the flute, or the clarinet. In addition, the bell is a wider opening, therefore the sideways spread of the sound is smaller and allows more directionality, especially for higher frequencies (see section 2.6). In other words, the volume and timbre perceived is very dependent on the direction the instrument is pointing.

---

[5] See table 5.2 for definition of intervals.

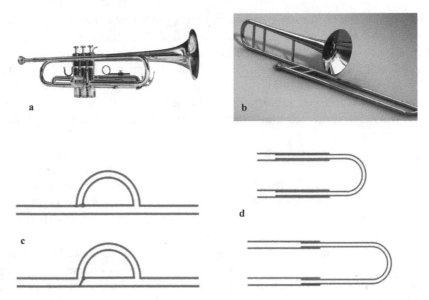

Figure 11.12. (a) Trumpet. Attribution: Amada44, CC BY-SA 3.0, via Wikimedia Commons. (b) Trombone. Attribution: Josef Plch, Public domain, via Wikimedia Commons. Schematic illustrating the principle of extending the effective length of an air column by (c) using a valve, and (d) using a sliding U-section.

11.5 String instruments

String instruments use stretched strings as vibrators. In contrast with percussion and wind instruments that can radiate the vibrational energy to the surrounding air efficiently enough to be heard, the transfer of energy from the string to the surrounding air is not efficient at all. This is because the contact area between a string and the surrounding air is small. The result is that the sound we hear is very weak, although plenty of vibrational energy is there. So, we need to have a path for the vibration energy to get out of the vibrating string and into the surrounding air. This is the function of the **radiator**, which is the body of the guitar or the violin. The vibrational energy is transferred to the front and back surfaces of the body, the so-called front and back plates, and to the air inside the body of the instrument.

The strings are made of different materials. Earlier instruments used catgut strings (made of sheep intestines). Modern instruments use metals, synthetic materials, or gut. As discussed in section 6.2, the mode frequencies of the standing waves supported by a string depend on the vibrating length of the string, and the speed of sound in the string. In turn, the speed of sound in the string depends on the linear density of the string, which is the mass per unit length of string, and on the tension on the string (see section 6.15 for details). Strings of higher linear density, are referred to as 'heavy' and lower linear density are called 'light' strings. Other things being equal, the heavy string produces a lower frequency than a light string. Also, other things being equal, the higher the tension on the string, the higher the frequency of the tone. The tension that must be applied on the string to produce the right tone depends on the size of the instrument and the type of the string.

Depending on the excitation mechanism, string instruments are further classified into *plucked* and *bowed* instruments, although this distinction is not rigid because bowed instruments can be both bowed and plucked. Bowed instruments include the violin, viola, cello, double bass and others. Plucked instruments include the ukulele, mandolin, banjo, lute, guitars and other. The guitars are further classified as classical, steel string acoustic and jazz guitars. We will first describe the parts and function of two of the most common string instruments, namely the guitar and the violin.

11.5.1 The guitar

The guitar does not belong to the string section of the symphony orchestra. The six string guitar dates back to the late 1700s. Today we have numerous types, with one neck, two necks, three necks, with frets, without frets, six strings, eight strings, and so on. Here we will describe the most common six string guitar.

Figure 11.13 shows the basic parts of a guitar. The strings are stretched between the **nut** and the **bridge**. Guitars usually have six single strings in six **courses**. The strings are resting on the **saddle** (white line in figure 11.13) affixed on top of the bridge. The distance between the nut and saddle defines the **scale length** of the instrument. For guitars the scale length is most commonly in the range of 0.61–0.66 m (24–26 in). The tension of the strings is adjusted using **tuning pegs**. The tuning pegs of the guitar are connected to a gear system (the tuning 'machine') that allows for fine adjustment of the string tension. The strings are named by the tone they produce when the entire length of the string between the nut and the saddle is vibrating. The string is then said to be *open*. The strings are also named by numbers (1st, 2nd, and so on) starting from the bottom string to the top string. The standard tuning of open strings for the guitar together with the note frequencies are listed in table 11.2.

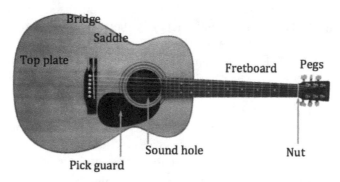

Figure 11.13. Basic parts of a guitar.

Table 11.2. Standard tuning of open strings for guitars.

String	E4 (1st)	B3 (2nd)	G3 (3rd)	D3 (4th)	A2 (5th)	E2 (6th)
Frequency (Hz)	329.6	246.9	196	146.8	110	82.4

The top-side of the arm, or **neck** of the guitar has metallic frets, forming the **fretboard**. The vibrating length of the strings can be changed by pressing, or *stopping* the string against the fretboard. On the guitar, the midpoint (usually marked by two inlays on the fretboard) is at the 12th fret from the nut. In other words, 12 steps on the fretboard correspond to one octave, meaning that the *frets* correspond to intervals of one *semitone*. As discussed in section 5.5 the interval of one semitone in the equal temperament tuning is 1.059 46, meaning that moving in steps of one fret from the nut towards the bridge increases the frequency by a factor of 1.059 46 or about 6%. Because the frequency is inversely related to the vibrating length (shorter length gives higher frequency) we conclude that on moving in steps of one fret towards the bridge, the vibrating length of the string *decreases by a factor* of 1.059 46. Let's assume that the open length of the string is 0.600 m, and calculate the vibrating length, that is the distance of the fret to the saddle, for the first three frets. For each step the distance decreases by a factor of 1.059 46, so:

1st 0.600/1.059 46 = 0.566 m; **2nd** 0.566/1.059 46 = 0.534 m; **3rd** 0.534/1.059 46 = 0.504 m.

The decrease corresponding to each step is:

0.600–0.566 = 0.034 m; 0.566–0.534 = 0.032 m; 0.534–0.504 = 0.030 m.

In other words, the frets are not equally spaced, and the spacing decreases as we move towards the bridge. As the frets are spaced based on the equal temperament scale, the guitar cannot play the just scale or quarter-tones. What is also interesting to note from table 11.2 is that interval between adjacent strings is five semitones (see figure 5.1), except that between the 3rd and 2nd string which is four semitones. Recall that in section 5.5 we called the five-semitone interval a fourth, and the three-semitone interval a major third. So, the guitar is tuned in fourths except for the 2nd to 3rd, which is a major third. This means that if we stop the sixth string at the fifth fret (second white dot of fretboard in figure 11.13) we have an A2, which is the same as the open fifth string. The same happens between all adjacent strings, except for going from the third to the second. Here the interval is a major third, and it is the 4th fret on the third string (between the first and second white dot in figure 11.13) that gives the B3. These unisons (meaning: same note) can be used as a quick way to check the tuning of the guitar.

The guitar **box** is made of carefully selected wood, usually spruce. The thickness of the **top plate** and **back plate** is about 2–3 mm (0.08–0.1 in) and both plates can have many vibrational modes, as discussed is section 11.3.2. The two plates are separated by the ribs. So, the standing wave frequencies that build up on the string can be transferred to the front and back plates of the box, and if the string frequencies match the mode frequencies of the box, we will have resonance. In addition, the box itself is an air cavity, that has its own resonance (see section 6.14.1). What is important is to have a distribution of resonances that will cover the playing range of the guitar. Note that the box does not add energy at all. The energy comes from the vibrating string, and is channelled to the box via the bridge. This is why touching a vibrating string causes the sound to stop at once. In terms of

11-17

radiated sound, the low frequencies come primarily from the top plate and the **sound hole**[6]. The back plate contributes more to the radiation of the high frequencies.

The guitar body has several braces on the inner side of the top plate, as well as the inner side of the back plate to provide some strength to the body, but also counteract the significant force from the tension of the strings. The total tension from the 6 strings can be about 100 lbs (about the weight of 45 kg). The tension tends to deform the top plate and bend the neck. The location of the braces is critical, because it affects the vibrational modes of the plates.

As mentioned in section 6.4 the timbre of the tone depends on the point where the string is excited. Plucking the string near the bridge emphasizes the higher harmonics, while plucking closer to the fingerboard produces a softer sound. The sound also depends on the direction of plucking. For example, if the string is plucked parallel to the top plate, the string vibrates parallel to the plate, the sound is softer and long-lived. If the string is plucked more vertically to the top plate, there is more vibration vertical to the plate, and faster transfer of the vibration energy from the bridge to the body of the guitar. As a result, the sound is louder and short-lived.

11.5.2 The violin

Figure 11.14 shows the basic parts of a violin. For the violin, the typical scale length, that is, the distance between the bridge and the nut, is about 0.33 m (13 in). The strings are stretched between the tuning pegs on the right, and the **tailpiece**, on the left. The tailpiece is attached to the **ribs**, which separate the **top** from the **bottom plate**. The top plate has two openings, the **f-holes**, shaped like the letter *f*. In addition to the tuning pegs, the violin can have 1 or 4 **fine-tuning pegs** on the tailpiece. The standard tuning of open strings for the violin together with the note frequencies are listed in table 11.3.

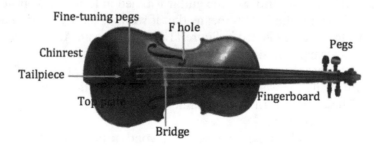

Figure 11.14. Basic parts of a violin.

Table 11.3. Standard tuning of open strings for violin.

String	E5 (1st)	A4 (2nd)	D4 (3rd)	G3 (4th)
Frequency (Hz)	659.3	440	293.7	196

---

[6] See video 6.3.

From figure 5.1, we see that the adjacent notes listed in table 11.3 are separated by seven semitones. The interval of seven semitones is called a *fifth* (see section 5.5) so, the violin is *tuned in fifths*. The strings are stopped by pressing them against the fingerboard. Note that there are no frets, which makes learning the violin more challenging. On the plus side, because there are no frets, the violin can play any frequency in its tonal range, meaning that the instrument can also play the *just scale*, and quarter-tones.

The thickness of the top plate and back plate is about 2–3 mm (0.08–0.1 in) and the vibrational energy transfers from the vibrating strings to the top plate through the bridge. A wood post, the **sound post**, connects the top plate to the back plate and transfers the vibrations to the back plate. The sound post is located under the bridge, on the side of the E5 string (1st string). The sound post also helps support the top plate which experiences a force of about 20 pounds (equivalent to the weight of a mass of 9 kg) that comes from the tension of the strings on the bridge. As an added support, the inner side of the top plate of the violin has a strip of wood that runs under the G3 string (4th string). This so-called **bass bar** not only provides strength to the top plate, but also plays a crucial role in the sound quality of the instrument by controlling the vibrational modes of the top plate. As in the guitar, the radiation of low frequencies is primarily from the top plate and the f-holes. The back plate contributes more to the radiation of the high frequencies.

Figure 11.15. The violin bow.

The violin bow is made of about 150 horsehairs stretched on a stiff stick. Quality bows are made of *Pernambuco* wood, although other types of wood and synthetic materials are common. The so-called **frog** (left end in figure 11.15) adjusts the tension of the bow. Excitation of the string by the bow occurs because of the friction that develops between the two. In addition, the horsehairs are rubbed with rosin (a material derived from pine trees) to make the bow sticky, and at least momentarily 'catch' the oscillating string, and transfer some energy to the string. This energy exchange depends on the speed of the bow and the force of the bow on the string. If the bow is moving too fast or too slow, or the force of the bow on the string is too high or too low, the resulting sound could be rather unpleasant. The force from the bow that causes the string to vibrate is not continuous. Instead, it follows a 'stick-and-slip' pattern, meaning that the bow momentarily catches the string, then the string slips-by the bow, then it catches again, and so on[7]. The rate at which the stick-and-slip occurs is determined by the frequency of the fundamental. In other words, a feedback process is established, as discussed earlier in connection with wind instruments.

---

[7] The stick-and slip is excitation at work in singing glasses (or glass harp) and the singing bowl in video 4.1.

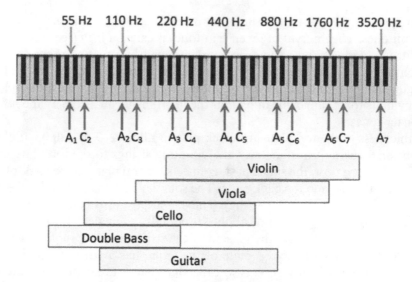

Figure 11.16. Playing ranges of some string instruments.

As mentioned in section 6.4, the timbre of the tone depends on the point where the string is excited. Thus, bowing the string near the bridge emphasizes the higher harmonics, and produces a louder and more brilliant sound. Bowing closer to the bridge requires higher and steadier force from the bow on the string. Bowing closer to the fingerboard produces a softer sound with less force.

11.5.3 Tonal range of some string instruments

Most acoustic guitars (as opposed to electric guitars discussed below) have six single strings, but there are many variations. So, some guitars have seven or eight single strings, some have six double strings, that is 12 strings in six courses, some have 12 double strings and so on. The scale length and tuning is the same for most types. For bowed string instruments we have a wide range of sizes: from the violin with scale length of about 0.33 m (13 in), to the double bass, with scale length of about 1.1 m (about 44 inches). Figure 11.16 shows the playing ranges of some string instruments. The piano keyboard is shown for reference. Note that the violin has the widest tonal range, almost four octaves. This is remarkable, considering the fact that the violin has two strings less than the guitar.

11.6 Further discussion

11.6.1 Electric versions of wind, string and percussion instruments

In section 7.4 we described the physical principles used in microphones. The same principles can be used for **pickups**, meaning devices that convert the vibrations produced by a musical instrument into an electrical signal. In section 7.8.8 we used the **electric guitar** as an example. The electric guitar does not have a radiator (sound box) and so, by design it outputs a very weak acoustic signal. Instead, it has a solid body, and produces an electrical signal that is fed to an amplifier. The electric violin is designed along the same lines. For electric guitars, the most common pickup is the magnetic, described in section 7.8.8. The main requirement for this kind of pickup is to have metal strings that contain iron or nickel, which interact strongly with the magnetic field of the pickup coils. This means that ordinary nylon or gut strings will not work with the magnetic pickup. The main objective of the magnetic pickup is to capture the motion of the string, so the magnetic coils need to be as *close to the strings* as possible, and *close to the anti-nodes* of the vibration works best. But as we learned in section 6.3, the nodes/anti-nodes of the overtones are in different places. In other words, we will need two, or three sets of pickups, at different spots, to capture as much of the overtone range as possible. The pickup closer to the bridge captures the higher frequencies, and the one closer to the neck picks up most of the lower frequencies. The contribution from each coil to the total output of the instrument can be adjusted by one to four *knobs*, which are essentially variable resistors. If the body is rigid, then the transfer of the vibrations from the bridge to the body is diminished and as a result the vibration of the string lasts longer. In other words, the envelope of the sound will have longer sustain and release stages (section 3.6) which makes the timbre of the electric guitar very distinguishable from the acoustic guitar. The magnetic pickups do not work as well for the violin, one of the main reasons being the lack of space to accommodate them. A common setup for violins is to use a saddle that is also a pick up.

Besides the purely electric guitar, and violin, there are acoustic instruments, that is, instruments with sound boxes, that come with factory installed pickups. These are sometimes referred to as semi-acoustic. Finally, there is a variety of pickups that can be attached to a purely acoustic string or wind instrument. Of course, the main concern here is to capture the timbre of the instrument with minimum disturbance, and without inconveniencing the player. In some cases, a good microphone is the best solution. A disadvantage of using microphones is that microphones tend to pick up sound from the loud speaker of the sound system, and pass it through the amplifier again and again, in other words, noise from a feedback loop, as described in section 7.8.11. Generally, the pickups for violins follow the lines of the main microphone types, that is dynamic, condenser (referred to as electrostatic) and piezoelectric, described in section 7.4.1. Some attach on the bridge, and some are fitted between the top-plate and the bass foot of the bridge. Some are in direct contact with the strings, in the part between the bridge and the tailpiece. Some clip-on to the top-plate, at the bass side f-hole, and other designs simply strap-on to the body, between the bridge and the fingerboard. Pickups for wind instruments operate on similar principles, except that they require direct contact with the air vibration inside the bore of the instrument. Usually this requires drilling a hole near the mouthpiece, and using a threaded fitting to screw-in the pickup.

When necessary, the sound of percussion instruments can be amplified using conventional microphones. What is more popular is the so-called **electronic drum** set, as opposed to electric. The electronic drum set does not have the basic vibrating elements, plates, membranes, rods, and the like. Instead, the set has a collection of sounds, mimicking the sounds of ordinary acoustic drums stored in memory. The drum set still consists of a number of components, more or less similar in shape to the components of an ordinary acoustic drum set, like a snare, a cymbal, and so on. Each component has a large number of sensors, or triggers, distributed over its surface. When one of the components, say the 'snare' is hit at any given point of the surface and with a specific speed, the triggers at that point of the surface send an electrical signal to the *module* or *brain* of the set. The module then retrieves from the memory the sound that corresponds to 'snare' hit at that particular point and at the specific speed. One major advantage of electronic drums is that they produce minimal sound, so the user can use a headphone set at home for quiet practice.

11.6.2 Overview of some string instruments

The violin and the guitar are relatively recent instruments that evolved from older forms, many of which are still in use. Probably the most common feature shared by the wide variety of string instruments is the need for a radiator. In one of the oldest forms, shown in figure 11.17(a) the radiator was a relatively thick piece of carved wood, without a top plate. This is common among the more traditional instruments, that is, to have the body carved out of wood, while some instruments use a gourd, like the *sitar*, or a turtle shell. Either way, most of the sound radiation has to come from a more flexible thin top plate, which can be from wood or animal skin, such as the West African *kora* shown in figure 11.17(b), the *banjo*, which traces its roots from West Africa, the *ehru*, which is a two-string Chinese fiddle, and many others.

One can also look at the way the strings are played. In some instruments the, strings are always open, like the *harp*, the *kanoon*, common in Arabic, Persian and Turkish music shown in figure 11.17(c), and the Japanese *koto*. Other instruments have fingerboards with frets, for instance, the *lute* and the *bouzouki* shown in figure 11.18(a) or without frets, for example, the *guqin* (Chinese zither) and the *oud*

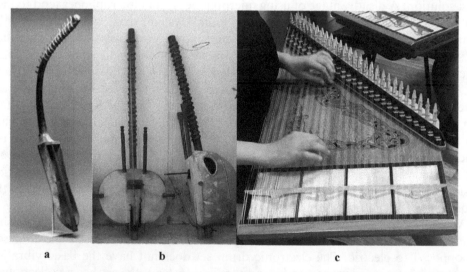

a　　　　　**b**　　　　　**c**

Figure 11.17. (a) Ancient Egyptian harp, ca. 1390–1295 B.C. Attribution: Metropolitan Museum of Art, CC0, via Wikimedia Commons. (b) Kora. Attribution: Mathaz, CC BY-SA 4.0, via Wikimedia Commons. (c) Kanoon or qanun. Attribution: Benoît Prieur—CC-BY-SA, via Wikimedia Commons.

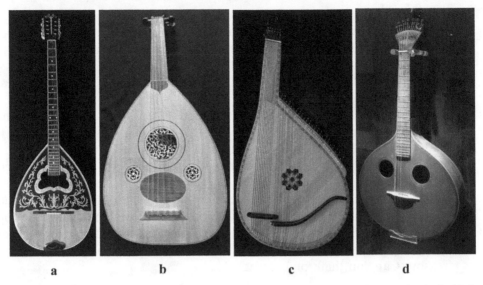

Figure 11.18. (a) Eight-string, four course bouzouki. (b) Eleven-string, six course (lowest string is single) Oud. (c) Ukrainian bandura. Attribution: Julianhayda, CC BY-SA 3.0, via Wikimedia Commons. (d) Portuguese guitar.

which is a lot like a fretless lute, and is shown in figure 11.18(b). One should keep in mind though, that the dividing lines are not so rigid. For example, the bandura, shown in figure 11.18(c) which has some strings over a fingerboard, and others that are played open. The metal frets in string instruments is a relatively new invention (around the mid-1800s). Frets were usually made from gut, tied on to the neck of the instrument, or bone, for instruments like the sitar and the *koto*. Unlike the metal frets that are fixed in position and are limited to the equal temperament scale, the gut and bone frets are movable, and so allow more flexibility in tuning. In most double stringed instruments of the guitar family, the top pairs combine a thick and a thin string, which are separated by one octave in pitch. The thicker string in most cases is the top one. Figure 11.18(c) shows one exception. The Portuguese guitar, or *fado* guitar, named so by the genre of music with which it is associated; the thinner string is the top one, which adds to its unique sound.

11.7 Problems and questions

1.
    (a) Which of the following instruments has a vibrator?
        Guitar, clarinet cymbal, drum.
    (b) For the above listed instruments, name the vibrator, if any.
2. Name the radiator for the instruments listed below:
    Guitar, clarinet, cymbal, drum.
3. Which of the following instruments has harmonic overtones? Explain your answer.
    Guitar, flute, cymbal, drum.

4.
    (a) What is the 'scale length' of a string instrument?

    (b) Which of the string instruments represented in figure 11.16 can play the lowest notes?

    (c) Which of the string instruments represented in figure 11.16 has largest scale length?

    (d) Which of the string instruments represented in figure 11.16 has the widest tonal range?

5. From the tonal ranges shown in figure 11.10 we note that the clarinet has lower fundamental pitch than the flute, although the flute is slightly *longer*. Explain the reason.

6. As mentioned in section 11.4, the total tension of the strings is significant and may deform the instrument. Would it be a good idea to replace the top strings, which are thicker and under higher tension, with thinner strings at much lower tension? The fact that this is not done tells us that there is a good reason. Can you think of a reason?

IOP Publishing

The Physics of Sound Waves (Second Edition)
Music, instruments, and sound equipment
Panos Photinos

Chapter 12

Keyboard instruments and the human voice

Console of the Wanamaker Grand Court Organ at Macy's Center City, in Philadelphia. Attribution: col_adamson, CC BY 2.0, via Wikimedia Commons

12.1 Introduction

In this final chapter, we will discuss some of the newest types of musical instruments, which are mostly instruments that use keyboards, and the oldest and most expressive musical instrument of all, the human voice. One can say that a keyboard is an interface between a human and a complex multicomponent device. For example, the keyboards of the Wanamaker pipe organ shown in the opening figure, allow the player to control 24 482 pipes. The design and construction of keyboards involves a fair amount of engineering, which, at least in the western world started in the Alexandrian period (around 300 BC). In the preceding, the so-called classical period, the great minds, like Plato and Aristotle, were concerned with pure science and ideas, and had no time or appreciation for anything practical or mechanical.

The first keyboard instruments were built around 1300, which is rather *recent*, compared to flutes and harps, which date back thousands of years ago. Although the pipe organ reached its maturity around the 1700s and the piano in the 1800s, the innovations continued. Small electric organs appeared in the 1930s and portable electronic keyboards that could emulate the sound of most musical instruments became available at affordable prices around 1980. Today a lot of music work is done by computer, and the keyboard is the most common interface for computer-based composition and performance of music. In this chapter, we will discuss the pipe organ, the piano and its predecessors, the synthesizer, and the human voice.

12.2 Pipe organ

The hydraulis[1] or water organ discovered in the 3rd century BC is the precursor of the pipe organ. While each pipe organ is different, most of the features found in pipe organs today, were more or less developed by the early 1600s. The basic idea is to use a mechanical system to blow air through a set of pipes. The system must produce a steady supply of high volume of air at low pressure to a large number of pipes.

Figure 12.1. The Wanamaker Grand Court Organ at Macy's Center City, in Philadelphia. Attribution: Zhicheng Zhao, CC BY-SA 4.0, via Wikimedia Commons.

---

[1] Υδραυλος, meaning water flute in Greek.

The typical air pressure is in the order of 700 Pa (about 0.1 psi) up to 25 000 Pa (about 4 psi). The space and structure required to accommodate the pipes, and the air supply system, the so-called **wind**, make the pipe organ an instrument more suitable for large venues, primarily churches and music halls. The larger organs can have well over 10 000 pipes, that can range in length from about a couple of centimeters (about 1 inch) to about 20 m (64 ft) in some cases. The pipes are organized in **ranks**, typically over one hundred, which are groups of pipes that simulate one specific instrument, for example, the flute, the strings, the trumpet, the clarinet and so on. In other words, the pipes within each rank have the same timbre. Figure 12.1 shows the Wanamaker organ, the largest pipe organ presently in operation, which has 28 482 pipes organized in 461 ranks.

12.2.1 The console

The instrument is controlled by one or more consoles, so the instrument can be played from different places. A console has anywhere from one to seven keyboards, the **manuals**, (see opening figure) as opposed to the **pedalboard**, shown in figure 12.2, which is operated by foot. The minimum number of manuals needed to play compositions from the baroque era is two. Three to five manuals are a typical number for large organs. The **compass** of each full manual is 61 keys, that is, five octaves, starting from C and ending in C. The compass of the pedalboard, shown in figure 12.2, is 32 pedals, to sound the lower notes (two and one-half octaves from C to G)[2].

The ranks are combined into groups that can be played by one of the manuals. These combinations are called **divisions**, and each division can be thought of as a separate organ. The divisions have names like 'great', 'swell', 'choir' and so on. Generally, there are more divisions than manuals, meaning that some divisions have to be **coupled** to a manual when needed. In addition to the manuals, the console has a number of **stops**, that control which rank is *speaking* and which rank is *stopped*. In other words, there is one stop for each rank, and the terms stop and rank are used interchangeably. The so-called **pistons**, are knobs that allow the player to activate a preselected combination of ranks all at once (light colored knobs under keys in chapter opening figure).

Figure 12.2. Pedalboard. Attribution: MoTabChoir01, Public domain, via Wikimedia Commons.

---

[2] The numbers follow the specifications of the American Guild of Organists.

Figure 12.3. Ranks of pipes on toe board (bottom of figure). Attribution: Xauxa Håkan Svensson, CC BY-SA 3.0, via Wikimedia.

The stops control which rank is active, and the manuals and pedalboard select which pipe or pipes within the selected rank is speaking. The activation of a pipe happens by mechanical linkages (the **trackers**) or magnetic valves inside the so-called **windchest**, which is where the wind is admitted to the pipes. Most of the pipes are set vertical with their lower end, the so-called **toe**, fitted in holes on the top surface of the windchests, the so-called toe board, shown in figure 12.3. The **slider** is a board with a pattern of holes, that can move horizontally underneath the toe board. When the player pulls a stop to activate a rank of pipes, essentially what happens is that the holes on the slider line up with the toe holes that belong to the selected rank, and wind can be fed into the pipes.

In terms of the overall amplitude of the sound, one should note that the speed of hitting the keys on the manual has no effect on the volume. Also, controlling the volume by feeding more wind to the pipes is not really an option, because this would also affect the timber of the sound, and in some cases may cause the pipes to vibrate in the second harmonic (see section 11.4.4). Actually, the wind system is designed to keep the wind as steady as possible. One option is to add or subtract ranks. The other option applies mainly to the swell division. Here the pipes are usually set in a box, open only on the side facing the audience. The sound from the box can be controlled by a system of vertical blinds activated by a special pedal on the console.

12.2.2 The pipes

Most of the pipes have circular cross section and are made of metal, usually a tin–lead alloy. Some pipes are made of wood, usually the longer ones, and have square or rectangular cross section. Depending on the excitation mechanism, we can have two types of pipes. If the pipe uses an air-reed (like the flute, see figure 11.8(a)) we have a **flue**-type pipe. The pipes in the background in figure 12.4 are flue-type. One can distinguish the flue type pipes from the characteristic horizontal slot, the mouth of the pipe (like the recorder in figure 11.8(b)). If the excitation happens by a vibrating reed, we have a **reed**-type pipe, like the clarinet (see figure 11.7(b)). The reeds in organ pipes are made of brass. Here there is a difference: in the case of the clarinet, the rate of vibration of the reed is set through a *feedback* process involving the vibration of the air column (section 11.4.1). In the organ reed pipe, the vibration of the reed is set by the vibrating length and stiffness of the reed. The length, and so the vibration frequency of the reed, is adjustable, and the tuner has to make sure that the length of the pipe, here called the **resonator**, matches the frequency of the vibrating reed. A selection of flue and reed pipes is shown in figure 12.5. The little wire at the bottom of reed pipes serves to tune the vibration frequency of the metal reed (see also bottom center of figure 12.4).

Tuning of the pipes is primarily done by adjusting the length of the vibrating air column. Some pipes have a tuning sleeve, that can be moved up or down to flatten or sharpen the pitch of the pipe. Similarly, for pipes closed in one end, the so-called **stopped** pipes, one can tune the pipe by moving the stop (which is essentially a plug inside the pipe) up or down[3]. In terms of ranks:

the *flue* type is used in the three main categories, namely *flutes*, *strings*, and *diapason*;

the *reed* type is used in the *clarinet*, *trumpet*, *horn*, *oboe* and others.

Each of the above categories has several sub-categories. For instance, in the strings we have *viole*, *violoncello*, and so on. Also, some stops control multiple ranks. So, these **mixture stops** have a number of ranks, indicated by a roman numeral, for instance III, or V, indicating that there are three or five ranks. Others have double pipes that are slightly off tune. Being off tune, they create beats, in other words, the intensity goes up and down, which is the so-called **amplitude vibrato** effect.

The function of the pipes follows what was described in sections 6.6 and 6.7, with some differences in the details, and the names. Recall from section 3.6 that the timbre of an instrument is determined by the amplitude spectrum which tells us about the distribution of intensity among the overtones, and the envelope which tells us how quickly the sound builds up and stops. There are several characteristics that affect the amplitude spectrum in pipe organs. For example, some of the organ pipes are open at both ends and some are closed at one end. As we know from section 6.6, open cylindrical pipes have all the harmonics (odd and even) while the pipes closed at one end have odd harmonics only. So, in the flue categories, the strings and diapason use open pipes. In the flute category we may have open or closed pipes. Also, the amplitude spectrum is affected by the geometric proportions of the pipe, that is, the ratio of the diameter to the length of the pipe. If the pipe is relatively wide

---

[3] The term *stopped* here has nothing to do with the stopping action from the console mentioned earlier.

Figure 12.4. Vertical flue pipes in cabinets. Horizontal reed pipes. Attribution: Frank Vincentz, CC BY-SA 3.0, via Wikimedia Commons.

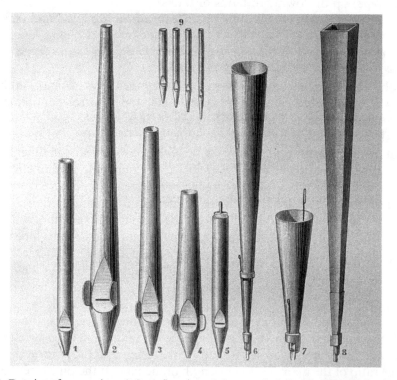

Figure 12.5. Drawing of organ pipes. 1–5 are flue pipes; 6–8 are reed pipes; 9 is a IV (four) mixture of flue pipes. Attribution: Biblioteca de la Facultad de Derecho y Ciencias del Trabajo Universidad de Sevilla, CC BY 2.0, via Wikimedia Commons.

in proportion to its length, the fundamental frequency of the pipe will be louder compared to the overtones. This is what happens in the flute category of flue pipes. If the pipe is relatively narrow, then the overtones will sound relatively louder, and this is the case of the string category in flue pipes.

The reed pipes can have different shapes, rectangular, cylinders, straight or tapered or flared, cones, and so on, and as discussed briefly in section 11.4.4, the shape may have significant effect on the spectrum of the overtones, which in some cases may not be harmonic. For example, in figure 12.5(6) the pipe sounds close to a trumpet and in figure 12.5(7) the pipe sounds like a human voice. Note also that the two pipes differ in the width of the **shallot**, which is the bottom end of the pipe against which the reed is flapping. The width of the shallot in relation to the length does affect the timbre, in the same way that the shape of the mouthpiece affects the sound of trumpets.

In terms of the envelope, as shown in figure 3.15, flutes have a relatively slow attack that leads smoothly to the sustain, followed by a fairly quick release. The sustain part can be as long as the player wants it to be; all it takes is to hold the key down. The quick release is compensated by the fact that the venues that house pipe organs, churches and large halls, usually have a significant reverberation time (section 9.4), which adds to the majestic sound of the instrument. The main concern with the envelope has to do with the wide tonal range of the instrument. It takes a few round trips up and down the pipe for a wave to reach the *sustain* level. For a short pipe, say 0.3 m (1 ft) long, ten round trips will take 0.017 s. For a 3 m (10 ft) pipe, it will take ten times longer, that is 0.17 s. So, the sound from pipes of different length does not reach its *fullness* together.

12.2.3 The pitch

The name of the rank or stop, tells us about the timbre of the sound, but it does not tell us anything about the frequency. For instance, we know that the leftmost key on the manual is a C, but which C is it? The middle C? And is it always[4] a C? The traditional way of specifying the pitch of a stop is to give a length of an open pipe in feet, say 8 ft (about 2.44 m). This means that the leftmost key on the particular manual sounds a frequency *approximately* equal to the fundamental frequency of an 8 ft open pipe. From our discussion in section 6.6, we know that the frequency of the fundamental is:

Frequency = 343/(2 × length of pipe in meters) = 343/(2 × 2.44) = 70 Hz,

which is close enough to C2 (= 65.4 Hz)[5]. This means that the manual starts with C2 on the left end, and that the highest note is C7. This pitch corresponds closely to where the C's are in the piano keyboard, that is, the middle C (C4) is near the center, closer to the left end. Similarly, a 16 ft stop will have C1 on the leftmost and C6 on the right end.

---

[4] See problem 5.
[5] See problem 4.

12.2.4 Pipe organs in entertainment

Because of their high volume of sound, pipe organs were used in other venues as well, besides concert halls and churches. Examples are, the Wanamaker pipe organ, originally built for the 1904 World's fair in St. Louis, and the Boardwalk Hall pipe organ in Atlantic City, which is the largest ever built, and currently under restoration. Modified pipe organs were also used in various shows, including the circus, and the silent movies. The theatre organs in particular added many new sounds (including cars, fire engines, and galloping horses!) as needed to complement the action on the screen. As a sound, the organ still remains an integral part in sports arenas, although the instruments used are mostly electrical or electronic.

12.3 Piano

The name *piano* comes from the instrument's ability to play both soft and loud, *piano e forte*, in Italian. *Piano e forte* soon became *pianoforte* (or *fortepiano*) and finally *piano*. The sound production in a piano is similar to the middle eastern *santur*. Like the western hammer dulcimer, the *santur* player uses a pair of sticks to strike a set of double strings. The major difference of course is that the piano has a keyboard, and a mechanical system, called **action**, to do the striking. The keyboard was first introduced in the 1300s, in connection with the pipe organ, and the idea of using a keyboard in connection with strings followed naturally.

12.3.1 Clavichord

The **clavichord** is the earliest keyboard instrument that uses strings. It is a small, portable instrument that appeared in the 1400s. Like the *santur*, the clavichord uses a pair of strings for each note. The keys are mounted on pivots, so that the far end lifts-up when the key is depressed. A small wedge-shaped piece of metal, usually brass, is attached at the far end of the lever, as shown in figure 12.6(a).

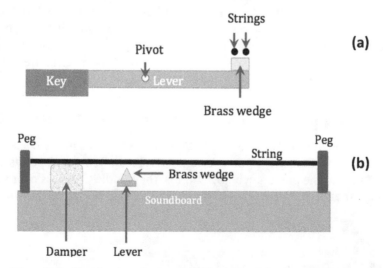

Figure 12.6. Clavichord mechanism. (a) Side view and (b) view from keyboard.

The undisturbed strings are resting on a damper (a piece of felt), and are clamped at both ends, as shown in figure 12.6(b). When a key is depressed, the brass wedge moves up, hits and rests on the strings. This action sets to vibration the section of the strings between the wedge and the peg on the right in 12.6(b). The damper keeps the section on the left from vibrating. So, in a way, the wedge acts as a fret in a guitar.

To accommodate more keys without adding more strings, more elaborate designs were invented where each pair of strings could be shared by three keys. The next step in the development of string-based keyboard instruments was to invent new ways to excite the strings, that is by plucking or by striking, which led to the harpsichord and to the piano.

12.3.2 The harpsichord

The **harpsichord** was invented in the early 1500s, and is very similar to the clavichord, except that each key has its own pair of vertically stacked strings, and that the strings are stretched perpendicular to the keyboard. This allowed for more strings, more keys, and more sound volume. The plucking is done by attaching a small pick, the **plectrum**, on a rod, the **jack**, which essentially replaces the brass plate in figure 12.6. The jack moves upward when the corresponding key is depressed. A major problem with both the clavichord and the harpsichord is that their sound is weak, and the player has limited control over the volume of the sound, in other words, the dynamic range is small. Another problem is the compass, which is basically determined by how many strings can be fit in the box. Keep in mind that the strings are under tension, which can be substantial. For instance, the tension of the six strings of a guitar can be the equivalent of 50 kg (110 lbs). So adding strings requires stronger braces, otherwise, the instrument could collapse! Finally, the response of the keyboard sets some limitations on what can be played on the instrument. The problem here is that the key must be released before it can be struck again. This puts a limit on how quickly a note can be repeated. Nevertheless, the clavichord and harpsichord have wonderful sound, with many compositions especially for their sound, which is more suitable for chamber music rather than symphony halls.

12.3.3 The parts of the piano

In the early versions of the piano, the pianoforte, the strings lay flat, and **hammers** made of felt or leather were used to strike the strings, a lot like the harpsichord. Later versions had a more intricate mechanism of **dampers**, shown in figure 12.7(a), usually made of felt. The dampers keep the strings of each note from vibrating while the note is not being played[6], and stop the vibration of the string once the corresponding key is released. In the later versions, the hammers, located under the strings, are activated by an **action** mechanism, that first lifts the damper off the strings, then sets the hammer to motion which trikes the strings. With this mechanism, the player can control the volume by controlling the speed and pressure

---

[6] That could come from sympathetic vibration, see footnote 4 in chapter 6.

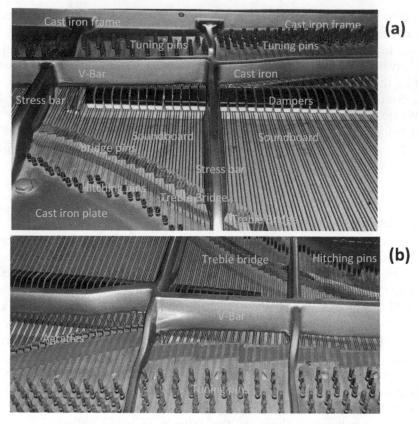

Figure 12.7. (a) View from tail. (b) View from keyboard side.

used to depress the keys. With time, more innovations were added, for example the use of pedals that mute the sound, or sustain the vibration, as described below.

The two most significant innovations that make the grand piano the unique instrument it is today were introduced in the early 1800s. The first significant innovation was the so-called *double pilot action*, which is a mechanism that allows the player to repeat a note without having to wait for the hammer to return to its starting position. In a good grand piano one can play a note about fifteen times in one second! The second significant innovation was the use of the cast iron frame. This means that we can have more strings and more octaves. The standard compass of a grand piano is 88 keys, with over 220 strings of various diameters (about 20 diameter sizes) with a total tension of over 20 tons. This is the weight of a fully loaded charter bus! The lowest note on a grand piano is A0 (27.5 Hz) on the left end of the keyboard. The highest note is C8 (4186 Hz) on the right end.

The strings are stretched between the tuning pins, which are on the keyboard end of the string, and the hitching pins, on the far side, as shown in figure 12.7(a). The speaking length is from about 0.1 m (4 inches) to over 2 m, and is defined by the distance between the bridge on the far end (the so-called **tail**), and the **agraffes** (for the heavier strings) or the **V-bar** (for strings on the treble side) on the keyboard side, as shown in figure 12.7(b). The agraffes are threaded brass posts with holes, that are screwed to the frame, and set the lateral spacing and the height of the strings. The V-bar[7] sits directly on the strings, and in a way, acts like a guitar fret. The bridge sits on the soundboard, and most grand pianos have two separate bridges, the bass bridge and the treble bridge, both shown in figure 12.8.

Figure 12.8. Treble and bass bridges.

Like in the guitar, the bass strings are wound, to reduce the effects of stiffness (section 6.14.2). The highest notes use triple strings, and the lowest use one string. In between there is a range that uses two strings. This is shown in figure 12.9.

The keyboard and action are sitting in the so-called **key bed**, which can be separated from the frame. Below the frame is the **soundboard**, that radiates the sound. Recall from our discussion of string instruments (section 11.5) that strings do not radiate sound efficiently, so as in the case of the guitar, the violin and other, the vibrational energy needs to transfer from the strings, through the bridge, to the body, from where it could be radiated more efficiently. The same happens with the piano. The sound-board is about 1 cm thick (about 3/8 inch) and is made of planks of wood (usually spruce) that are glued together, running roughly parallel to the direction of the bridges, which is more or less at an angle 45 degrees to the keyboard. Each plank is about 15 cm (6 inch) wide. About 10 wood ribs hold the planks together on the bottom side. The ribs add strength to the soundboard, but also affect the vibrational modes, in other words, the sound quality, just like the top and bottom plates of the guitar.

(a) **(b)**

Figure 12.9. Strings can be single (a) or double (a) and (b) or triple (b). Diameter of coin is 17.9 mm (0.7 in).

Most grand pianos have three **pedals**, as shown in figure 12.10. The function of pedals is usually as follows. The left pedal is the *soft* or *una corda* (meaning one string). It usually shifts the action slightly to the left so that the hammers over the multiple strings miss one of the strings. The middle pedal is the *sostenuto* which once depressed keeps the dampers of the keys that are already lifted from going back on

---

[7] Also referred to as *capo tasto* or *capo d' astro*.

Figure 12.10. Piano pedals for Baldwin baby grand piano. The number and function of pedals is different for different pianos.

the strings. The right pedal is the *sustain*, which lifts all the dampers, making the sound much fuller, or chaotic if overused.

The above discussion focused on the grand piano, which in terms of length could be from 1.5 m (5 ft) to over 2.8 m (over 9 ft), and can weigh over 500 kg (1100 lbs). Smaller types of pianos, the so-called uprights, that are more suitable for limited spaces, have become very popular over the years. The size is reduced by having the frame vertical rather than horizontal. The smallest are the spinets, which usually weigh under 130 kg (300 lbs). The larger uprights can weigh up to 450 kg (1000 lbs).

12.4 Human voice

Figure 12.11 shows the tonal range of human voices. The ranges refer to what is often called the **chest voice**. The frequencies refer to the fundamental. Note that on average each type of voice covers about two octaves, in other words, the frequencies at the two ends of a given voice, say the tenor, changes by a factor of four. The lowest frequencies are in the order of 100–200 Hz. The highest harmonic frequencies that can be produced by the human voice are about 5 kHz.

We know that one can sing an 'aa' or an 'ee' or an 'oo' on a C note, or any other note. Which tells us that we can sing different sounds on the same fundamental frequency. So, the human voice must include a number of harmonics, that can be modified by the singer, in order to make the 'aa' sound rather than the 'ee' or the 'oo'. Also, in singing, we mostly use vowels. One cannot carry a tune with 'ssss' or 'ffff' or 'tttt.' These sounds contain a very large number of frequencies, and have no recognizable pitch. These sounds are produced by the teeth or the lips, and all that is needed is a stream of air. One can carry a tune by whistling, which is an almost pure tone of frequency around 1000 Hz or so, which is at the very high-end of the tonal ranges shown in figure 12.11. This tells us that mechanism involved in whistling is different than what produces the chest voice. Whistling is related more to the way pipes produce sound. Singing vowel sounds does not work the same way.

Figure 12.12(a) shows the amplitude spectrum of a male voice singing an 'eee' as in 'cheese.' We note that there are some prominent and regularly spaced peaks below 500 Hz. The peaks are overtones of the fundamental, which is 100 Hz in this case. Note also the presence of a number of peaks around 2–2.5 kHz. Figure 12.12(b) shows the amplitude spectrum of male voice singing 'ooo' as in 'glow.' Here the fundamental is also 100 Hz, and there are several regularly spaced overtones up to 1 kHz, but in

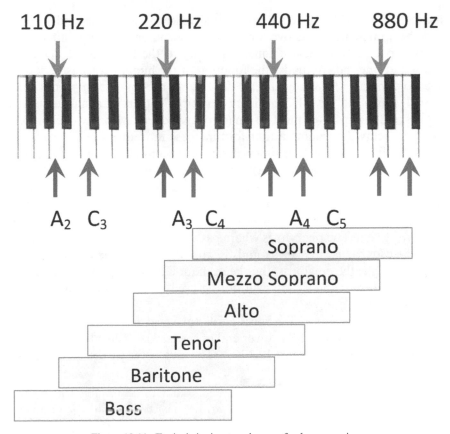

Figure 12.11. Typical singing tonal range for human voice.

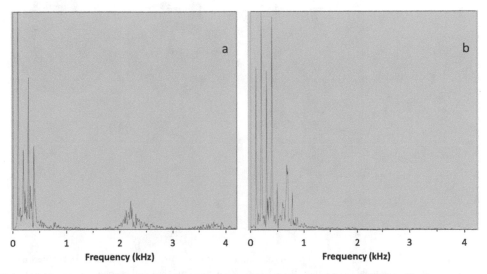

Figure 12.12. (a) Amplitude spectrum of male voice singing 'eee' as in 'ch<u>ee</u>se.' (b) Amplitude spectrum of male voice singing 'ooo' as in 'gl<u>ow</u>.'.

contrast to figure 12.12(a) there are no other significant features beyond 1 kHz. Note also that the amplitudes of the overtones follow an entirely different pattern.

The two spectra may suggest some similarity to the behavior of open pipes or pipes closed at one end. Actually, the function of the human voice is very different. For one, to generate a fundamental of 100 Hz, we need a 0.85 m (about 3 ft) long pipe closed at one end, or a 1.7 m (6 ft) open pipe. The entire distance between the vocal cords and the lips, is at most 20 cm (about 8 inches). Also, the chest voice has a range of two octaves, meaning a four-fold increase in frequency. In terms of pipes, increasing the frequency by four times, means shortening the length of the pipe by a factor of four, which is not what happens to our throat when singing. To describe the mechanism involved in chest voice, it is helpful to identify the parts involved.

The entire 'instrument' or **vocal tract** consists of the **larynx** (the **voice box**), the **nasal cavity** and the **mouth**, as shown in figure 12.13(a), and shares pathways with the digestive system. The pharynx is where pathways cross: it connects to the mouth, the nose, the esophagus, the larynx and the middle ear! As one might expect, there are safeguards, like the epiglottis, that keep the solids and liquids from going down the windpipe. Most of these parts will not be used in our description, and will not be discussed further.

The source of vibration is the **vocal cords** (or **vocal folds**) which are essentially muscle membranes located at the lower end of the larynx. When breathing, the folds are retracted, as shown in figure 12.13(b). The gap between the vocal folds is the

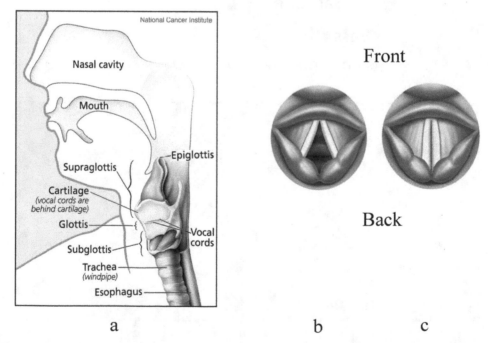

Figure 12.13. (a) The main parts of the larynx (supraglottis, glottis, and subglottis) and other nearby structures, including the nasal cavity, mouth, cartilage, vocal cords, trachea and esophagus. Attribution: Alan Hoofring (Illustrator), Public domain, via Wikimedia Commons. Top view of vocal folds. (b) When breathing (c) When speaking. Attribution: Prejun, CC0, via Wikimedia Commons.

glottis. Figure 12.13(c) shows the arrangement of the folds when vocalizing (speaking or singing). We see that the air pathway is restricted, and this is why it is hard to talk or sing when running. The air supply comes from the lungs, which act like a reservoir of air that inflates and deflates as we breathe. The lungs do not have muscles, instead the air is drawn-in and pushed-out by the action of the diaphragm or the expansion of the thorax (the rib cage).

As an instrument, the human voice is in a way similar to the trumpet, in that the glottis injects puffs of air into the vocal tract. The major difference is that in a trumpet the feedback mechanism (see section 11.4.1) determines the rate at which the lips are pushed open. To keep things simple, we can approximate the puffs as a square wave form, analyzed in section 3.10.3[8]. The spectrum of the puff has many frequency components with diminishing amplitudes, as shown in figure 3.18. From the spectra of figure 12.12, we see that the amplitudes are not diminishing uniformly, in fact in figure 12.12(b) the amplitude of the third overtone is larger than the fundamental. This means that some 'filtering' of frequencies occurs in the vocal tract. This 'filtering' must be adjustable as well, and that would explain why we have higher frequency peaks in figure 12.12(a) and not in figure 12.12(b).

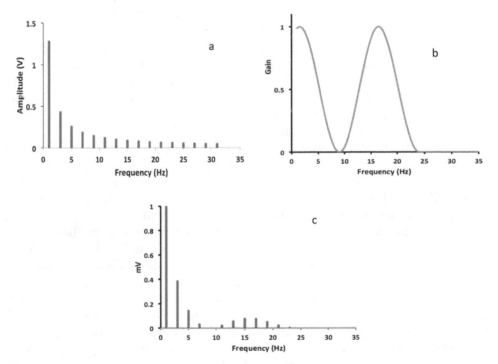

Figure 12.14. (a) Amplitude spectrum of square wave form, representing the 'puff.' (b) Gain characteristics of filter. (c) Amplitude spectrum of the output of the filter, which is the product of graph (a) times graph (b).

Figure 12.14 shows an oversimplified model of the action of a filter. The input to the filter is the amplitude spectrum of the square waveform, figure 12.14(a), and the

---

[8] The fact that the spectrum of the wave analysed in section 3.10.3 has odd frequencies only is not important.

gain characteristic of the filter is shown in figure 12.14(b). The output of the filter is the product of the input spectrum multiplied times the gain of the filter, and is shown in figure 12.14(c). Now the pharynx, the oral and nasal cavities are three-dimensional, and as described in section 9.11.3, three-dimensional spaces have a lot more resonance frequencies that depend on all three dimensions of the volume. In other words, by adjusting the lateral dimensions of the pharynx and the cavities, not necessarily the length, the singer adjusts the shape of the gain characteristics.

The fundamental frequency is set by the repetition rate of the puffs. What opens the glottis is the air pressure from the lungs. What closes the glottis is essentially a low pressure created by the fast-moving air stream, a lot like the action of the reed in the clarinet (see section 11.4.1). The muscles in the vocal folds contribute to the tension of the folds, which determines the rate of the puffs. The overtone amplitudes are adjusted mainly by changing the lateral dimensions of the shape of the pharynx and the oral and nasal cavities.

12.5 Synthesizers

In chapter 3 we discussed how to get the frequency spectrum of a sound. We also compared the frequency spectra of several instruments, and the rate at which the sound builds up and dies out. Having all this information about an instrument, say a classical guitar, one may reverse the process and create the sound of a classical guitar by putting together, or **synthesizing**, the sound from its frequency components. In a way, the progress that led to modern electronic synthesizers is parallel to the development of sound recording and reproduction discussed in chapter 10. In both cases, the early stages involved large and expensive devices, and in the modern version, the devices are compact and very affordable.

As with musical instruments, the essential part for making a sound is a vibrator that can produce a range of fundamental frequencies and overtones. In a synthesizer, the frequencies are produced by electronic **oscillators**, which are devices that produce voltages of selectable frequency. One can have a large number of such oscillators, one to produce the fundamental, and many others to produce the overtones. This method is called **additive synthesis**. One could also have oscillators that produce complex tones of selectable fundamentals. The complex tone is selected to have a wide range of overtones. The next step is to use filters (see section 7.3) to pass the frequencies needed for producing the desired tone and block the unneeded frequencies. This method is **subtractive synthesis**. Either way, additive or subtractive, the result is the amplitude spectrum of a musical sound. The next step is to make this sound build up and die out the same way the musical instrument does. This is what we called the **envelope** in section 3.6. The envelope consists of attack, decay, sustain, and release (ADSR), as shown in figure 3.13. For example, to produce a guitar-like sound, the attack must be quick and pronounced; to produce a flute-like sound, the attack must lead gradually to the sustain part. This can be achieved by amplifying the musical sound at the proper rate, and is done by the so-called **envelope generator**. To produce tonal effects such as vibrato (section 12.2.2) a separate low frequency oscillator is used to *modulate* the amplitude (amplitude vibrato) or frequency (frequency vibrato) of the musical sound.

Clearly all these adjustments cannot be done manually while trying to play a piece of music at the same time. In modern synthesizers, all the tasks described above, adding or subtracting frequencies, filtering, and the like, are voltage controlled. For example, the keyboard of the synthesizer produces voltages. Ascending by one key increases the voltage output of the keyboard. This voltage increase is electronically translated to a frequency increase in the fundamental of the oscillator. The increments in the voltage from the keyboard are selected such as to increase the fundamental of the oscillator by one semitone. Similarly, a switch on the synthesizer can select the envelope, the ADSR mentioned above. By selecting an ADSR, one can use the synthesizer to mimic sounds of known instruments and can also produce new sounds that do not correspond to the sound of any musical instrument. There are other manipulations that go well beyond what one can do with most instruments. For example, the sound from each violin in an orchestra is slightly different, because each violin is slightly different, and the pitch of each violin is not exactly the same. The result is that the sound from the violin section is not the same as one violin playing louder. This is the **chorus effect**. A synthesizer can produce a number of replicas that are slightly different, add them together and produce the chorus effect.

One can produce and store a musical piece using a synthesizer. Stored pieces of music can be exported or imported to and from other digital devices. Devices can exchange commands or execute commands that allow one digital unit to control what information is exchanged and when to do so. To enable this exchange, it is necessary to have a communication protocol, that is, a common language that all digital instruments and computers can understand. One of the most widely used protocols is the Musical Instrument Digital Interface, or **MIDI** for short, which is described in section 8.7.2. The MIDI can be a separate unit within a keyboard that allows the user to enter commands; for example, to tell a synthesizer what to play, how loud, and when. It could also be in the form of software installed on a computer, a so-called **digital audio workstation**, DAW for short.

The increased availability of digital memory allows one to store the sounds of musical instruments, or entire sequences of musical sounds, in digital form. The so-called **sampler** is essentially a library of sounds of different instruments. Using the instrument sounds stored in the sampler eliminates the need to synthesize these sounds. The sampler can be a separate unit that stores the sounds in its memory, or the sound library can be stored in computer memory. Either way, using commonly available software one can use the computer as a synthesizer. Such software allow the user to record and edit sounds, like change the pitch, the duration and timing of each note, or the entire piece. In addition, by creating different tracks for different instruments one can effectively create an entire music band.

12.6 Problems and questions

1. (a) Which of the following instruments has a vibrator?
        Pipe organ   Piano   Synthesizer   Human voice.
    (b) For the above listed instruments, name the vibrator, if any.

2. Name the resonator for the instruments listed below:
   Pipe organ   Piano   Human voice.
3. Many of us had the opportunity to observe a science demo where one produces a 'cartoon' voice by inhaling helium gas. The speed of sound in helium is about 1000 m s^{-1}. You may want to look at problem 8 in chapter 6 first.
   (a) Do you expect the vibrational frequency of the vocal folds to change by the inhaled helium?
   (b) What causes the high pitch in the cartoon voice?
4. In section 12.2.3, we found the fundamental frequency of an 8-ft (2.44 m) pipe:

Frequency = 343/(2 × length of pipe in meters) = 343/(2 × 2.44) = 70 Hz.

This value is 7% off the C2 (= 65.4 Hz). The pipes in an organ are real pipes (section 6.14.3). Can you think of a reason for the discrepancy?
5. We have two open pipes. Pipe A is 4 m long and pipe B is 2.666 (or 2 and 2/3) m long.
   (a) Find the fundamental frequency of pipe A.
   (b) Find the fundamental frequency of pipe B.
   (c) Find the ratio of the fundamental frequency of pipe B divided by the fundamental frequency of pipe A.
   (d) How does the fundamental frequency of pipe B relate to the harmonics of pipe A?
   (e) Find the ratio of the length of pipe A, divided by the ratio of pipe B. Is it close to your answer in part (c)? Explain why or why not.
   (f) Which interval corresponds to the ratio of part (c)?
   (g) Suppose that length ratio of pipe X divided by the length of pipe Y is the same as the length ratio of part (e). If the length of pipe X is selected to produce the note C4, what note will pipe Y produce?
   (h) Do the results of parts (c) and (e) remain the same if we measure the lengths of the pipes in feet rather than meters?
   (i) The pitch of organ pipes is not always an integer number of feet. The so-called **mutations** have fractional pitch, like 2 and 2/3 ft. What would be the note of the leftmost key on the manual for this rank?
   (j) Suppose that we sound a rank of 4 ft and a rank of 2 and 2/3 together, what would be the result?

Bibliography

The following seven books are among my favorite in terms of pedagogical approach, and require minimal mathematical background:

Backus J 1977 *The Acoustical Foundations of Music* 2nd edn (New York: W W Norton)

Berg R E and Stork D G 2005 *The Physics of Sound* 3rd edn (Englewood Cliffs, NJ: Prentice Hall)

Campbell M and Greated C 1994 *The Musician's Guide to Acoustics* (Oxford: Oxford University Press)

Johnston I D 2002 *Measured Tones: The Interplay of Physics and Music* 2nd edn (Bristol: IOP)

Moravcsik M J 1987 *Musical Sound: An Introduction to the Physics of Music* (New York: Paragon)

Rigden J S 1985 *Physics and the Sound of Music* 2nd edn (New York: Wiley)

White H E and White D H 2014 *Physics and Music: The Science of Musical Sound* (New York: Dover Publications)

For a discussion on psychoacoustics: Roederer J G 1995 *The Physics and Psychophysics of Music: An Introduction* 3rd edn (Berlin: Springer)

A very comprehensive book for science majors is: Rossing T D, Moore F R and Wheeler P A 2002 *The Science of Sound* 3rd edn (Reading, MA: Addison Wesley)

More advanced texts: Filippi P, Habault D, Lefebvre J-P and Bergassoli A 1999 *Acoustics: Basic Physics, Theory, and Methods* (San Diego, CA: Academic)

Loy D G 2006 *Musimathics: The Mathematical Foundations of Music* (Cambridge, MA: MIT Press)

A most rigorous mathematical description of musical instruments can be found in: Fletcher N H and Rossing T D 2005 *The Physics of Musical Instruments* 2nd edn (Berlin: Springer)

Two reliable websites of interest are: A reference for definitions: https://acoustic-glossary.co.uk/sound-pressure.htm

On musical instruments and demonstrations: https://newt.phys.unsw.edu.au/music/

IOP Publishing

The Physics of Sound Waves (Second Edition)
Music, instruments, and sound equipment
Panos Photinos

Appendix A

Scientific notation and abbreviations

1	10^0
1000	10^3 kilo (k)
1 000 000	10^6 mega (M)
1 000 000 000	10^9 giga (G)
0.001	10^{-3} milli (m)
0.000 001	10^{-6} micro (μ)
0.000 000 001	10^{-9} nano (n)

Units International units (SI) used:

m	meter (length)
s	second (time)
kg	kilogram (mass)
N	newton (force)
J	joule (energy)
W	Watt (power)
Pa	pascal (pressure)
Hz	hertz (frequency)
dB	decibel (dimensionless)
Ω	ohm (resistance or impedance)
V	volt (voltage)
A	ampere (electric current)

Appendix B

B.1 Symbols

The symbols used here in connection with waves and oscillations are listed below, with corresponding SI units in parentheses:

c = speed (m s^{-1})
λ = wavelength (Greek letter lambda, m)
f = frequency (Hz)
A_o = amplitude (Pa for a sound; m for strings)
x = distance coordinate (m)
t = time (s)
F = tension of spring (N)
d = linear density (kg m^{-1})

B.2 Traveling waves

Equation for wave traveling in the positive x direction:

$$A = A_o \cos(2\pi x/\lambda - 2\pi f t) \quad \text{or} \tag{B.1}$$

$$A = A_o \sin(2\pi x/\lambda - 2\pi f t) \text{ and more generally} \tag{B.2}$$

$$A = A_o \sin(2\pi x/\lambda - 2\pi f t + \phi). \tag{B.3}$$

Here ϕ is the initial phase. If the wave is traveling in the negative x direction, then the negative sign preceding the quantity $2\pi f$ becomes positive. Here the quantity A can be pressure for sound waves or displacement for water waves or strings. A_o is the amplitude of the wave. The above equations give the value of A at distance x away from the origin, and at time t. Note that choice of the origin, and the instant when we started measuring the time, are arbitrary. Recalling that $\cos(0) = 1$, we note from equation (B.1) that if $2\pi x/\lambda - 2\pi ft = 0$, then $A = A_o$, that is, we have a crest. The crest moves with speed $c = x/t$. Since $2\pi ft - 2\pi x/\lambda = 0$, we find $x/t = f\lambda$, or

$$c = f\lambda. \tag{B.4}$$

This is the *phase velocity* of the wave.

B.3 Beats

We will use the trigonometric identity

$$\sin(a) + \sin(b) = 2\sin\left(\frac{a+b}{2}\right)\cos\left(\frac{a-b}{2}\right) \tag{B.5}$$

to add two oscillations of equal amplitude V and of frequencies f_1 and f_2

$$V\sin(2\pi f_1 t) + V\sin(2\pi f_2 t) = 2V\cos\left(2\pi\frac{f_2-f_1}{2}t\right)\sin\left(2\pi\frac{f_1+f_2}{2}t\right) \tag{B.6}$$

We see that the second factor on the right-hand side is an oscillation with frequency equal to the average of f_1 and f_2. The frequency of the *cos* term is equal to half the difference between f_1 and f_2.

We will apply this result to the example of figure 1.8 and choose $V = 1/2$. The corresponding frequencies of the individual waves are $f_1 = 0.1333$ Hz and $f_2 = 0.1666$ Hz. The average of f_1 and f_2 is 0.15 Hz. One half of their difference is 0.0166 Hz, and equation (B.6) becomes:

$$\cos(2\pi 0.0166t)\sin(2\pi 0.15t) \tag{B.7}$$

The results are shown in figure B.1, where A indicates the sin term and B is the cos term in equation (B.7). From the figure we note that the effect of the negative cycle of the cos term is to 'flip' the values of the sin term. In other words, the beat pattern repeats with a frequency of $f_1 - f_2 = 0.0333$ Hz rather than $(f_1 - f_2)/2 = 0.0166$ Hz.

We can apply equation (B.6), to find the repeat frequency of the envelope for second order beats (section 5.11.4.) Suppose that we are combining two frequencies

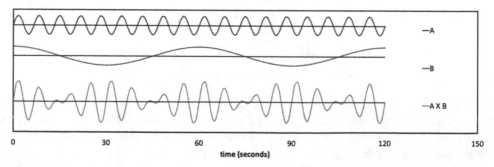

Figure B.1. Sum of two waveforms of frequencies 0.1333 and 0.1666 Hz. The green curve is identical to the green curve in figure 1.8.

f_2 and f_1 that happen to be close to a ratio of two small integers, n and m. So, we can write $f_2 = (n/m) f_1 + d$. Since $m f_2 = n f_1$, we find that the frequency difference in equation (B.6) is:

$$f_2 - f_1 = md.$$

For instance, if f_2 is off the octave ($n = 2$, $m = 1$) by 5 Hz, (meaning $d = 5$), then the repeat frequency of the envelope is 5 Hz. If f_2 is 5 Hz off the perfect fifth ($n = 3$, $m = 2$) then repeat frequency of the envelope is $md = 2 \times 5 = 10$ Hz, as found in section 5.11.4.

B.4 Standing waves in a string

The equation for standing waves is

$$A = A_o \cos(2\pi f t) \sin(2\pi x/\lambda). \tag{B.8}$$

Here A is the displacement from the equilibrium position of the string. The string is clamped at both ends, therefore, $A = 0$ at $x = 0$ and at $x = L$, where L is the length of the string. Recall that $\sin(n\pi) = 0$ if n is a positive integer (1, 2, 3, ...). For $A = 0$ at $x = L$ we must have $2\pi L/\lambda = n\pi$ or

$$\lambda = 2L/n, \text{ where } n = 1, 2, 3, \text{ etc.} \tag{B.9}$$

For $n = 1$ we have $\lambda = 2L$ (the fundamental mode), for $n = 2$ we have $\lambda = L$, for $n = 3$ we have $\lambda = 2L/3$ and so on for higher values on n, in agreement with figures 6.1–6.3. The phase velocity of a wave in a string is given in terms of the tension of the string F and its linear density d (which is the mass of the string divided by the length if the string)

$$c = \sqrt{F/d} \tag{B.10}$$

which must equal the product of the frequency times the wavelength. The latter is determined by equation (B.6) therefore

$$f\lambda = f 2L/n = \sqrt{F/d} \text{ or} \tag{B.11}$$

$$f = \frac{n}{2L}\sqrt{F/d} \tag{B.12}$$

It follows from equation (B.12) that:
- The fundamental mode ($n = 1$) has the lowest frequency.
- Longer strings produce lower frequencies.
- The frequency increases with tension.
- The frequency decreases if the linear density is higher, which means that thicker strings (of the same material) produce lower frequencies.

The phase velocity given by equation (B.4) does not refer to the speed of the up and down oscillation at each point along the string. The speed of oscillation V at each point is

$$V = -2\pi f A_o \sin(2\pi f t) \sin(2\pi x/\lambda). \tag{B.13}$$

We see that the oscillation speed is different at different points along the string, and also changes with time. The phase velocity refers to the speed of a single wave propagating in a very long string, as discussed in section 6.2.

Here we will show that standing waves in strings (and pipes) are formed by reflection of waves at the endpoints.

Consider a traveling wave

$$A = (A_o/2) \sin(2\pi x/\lambda - 2\pi ft) \tag{B.14}$$

and its reflection (i.e. traveling in the opposite direction, as indicated by the + sign preceding the quantity $2\pi ft$)

$$A = (A_o/2) \sin(2\pi x/\lambda + 2\pi ft) \tag{B.15}$$

(note that here the waves have amplitude $A_o/2$). Adding the two waves and using equation (B.5) we recover the standing wave equation (B.8).

The Physics of Sound Waves (Second Edition)
Music, instruments, and sound equipment
Panos Photinos

Appendix C

Note frequencies in equal temperament tuning

No note			Fundamental Frequency (Hz)
A0			27.50
A0♯	or	B0 ♭	29.14
B0			30.87
C1			32.70
C1♯	or	D1 ♭	34.65
D1			36.71
D1♯	or	E1 ♭	38.89
E1			41.20
F!			43.65
F1♯	or	G1 ♭	46.25
G1			49.00
G1♯	or	A1 ♭	51.91
A1			55.00
A1♯	or	B1 ♭	58.27
B1			61.74
C2			65.41
C2♯	or	D2 ♭	69.30
D2			73.42
D2♯	or	E2 ♭	77.78
E2			82.41
F2			87.31
F2♯	or	G2 ♭	92.50
G2			98.00
G2♯	or	A2 ♭	103.83
A2			110.00

(Continued)

(*Continued*)

No note			Fundamental Frequency (Hz)
A2♯	or	B2♭	116.54
B2			123.47
C3			130.81
C3♯	or	D3♭	138.59
D3			146.83
D3♯	or	E3♭	155.56
E3			164.81
F3			174.61
F3♯	or	G3♭	185.00
G3			196.00
G3♯	or	A3♭	207.65
A3			220.00
A3♯	or	B3♭	233.08
B3			246.94
C4			261.63
C4♯	or	D4♭	277.18
D4			293.66
D4♯	or	E4♭	311.13
E4			329.63
F4			349.23
F4♯	or	G4♭	369.99
G4			392.00
G4♯	or	A4♭	415.30
A4			440.00
A4♯	or	B4♭	466.16
B4			493.88
C5			523.25
C5♯	or	D5♭	554.37
D5			587.33
D5♯	or	E5♭	622.25
E5			659.26
F5			698.46
F5♯	or	G5♭	739.99
G5			783.99
G5♯	or	A5♭	830.61
A5			880.00
A5♯	or	B5♭	932.33
B5			987.77
C6			1046.50
C6♯	or	D6♭	1108.73
D6			1174.66
D6♯	or	E6♭	1244.51

E6			1318.51
F6			1396.91
F6♯	or	G6 ♭	1479.98
G6			1567.98
G6♯	or	A6 ♭	1661.22
A6			1760.00
A6♯	or	B6 ♭	1864.65
B6			1975.53
C7			2093.00
C7♯	or	D7 ♭	2217.46
D7			2349.32
D7♯	or	E7 ♭	2489.02
E7			2637.02
F7			2793.83
F7♯	or	G7 ♭	2959.95
G7			3135.96
G7♯	or	A7 ♭	3322.44
A7			3520.00
A7♯	or	B7 ♭	3729.31
B7			3951.07
C8			4186.01

Appendix D

Intervals of the Pythagorean tuning

Here we will use the Pythagorean tuning to assign frequency ratios for the C-major scale, and so derive the values listed in table 5.3. Recall that the tuning is based on the perfect fifth. Ascending by a fifth means moving seven semitone steps to the right. In terms of ratio, ascending by a fifth means multiply by (3:2). Descending by a fifth means seven semitones to the left, and in terms of ratio, it means divide by (3:2) or multiply by (2:3).

Start from a C4	Interval = 1 (unison)
C4 + fifth (7 semitones to the right) = G4	× (3:2)
G4 + fifth (7 semitones to the right) = D5	× (3:2) × (3:2) = (9:4)
D5 + fifth (7 semitones to the right) = A5	× (3:2) × (3:2) × (3:2) = (27:8)
A5 + fifth (7 semitones to the right) = E6	× (3:2) × (3:2) × (3:2) × (3:2) = (81:16)
E6 + fifth (7 semitones to the right) = B6	× (3:2) × (3:2) × (3:2) × (3:2) × (3:2) = (243:32)

So far, we have all the notes except F.
To get an F note, we start from C4 and ascend a fifth:

C4 − fifth (7 semitones to the left) = F3	× (2:3)

So, we managed to generate six notes. Now D5 and A5 need to descend by one octave, to become D4 and A4. To descend by one octave, we divide the corresponding ratios by 2.

Also, E6 and B6 need to descend by two octaves, to become E4 and B4. To do so, we divide the corresponding ratios by 4. F3 needs to ascend by one octave to become F4, so we need to multiply the ratio by 2.

The resulting ratios are:
C4 = 1
D4 = (9:4)/2 = (9:8)
E4 = (81:16)/4 = (81:64)
F4 = (2:3) × 2 = (4:3)
G4 = (3:2)
A4 = (27:8)/2 = (27/16)
B4 = (243:32)/4 = (243:128)

Which are the values listed in table 5.3.

Appendix E

Answers to problems and questions

Chapter 1

1.

(a) 120 beats per minute means 2 beats in 1 s, or 0.5 seconds per beat.

(b) 2 beats per second, or 2 Hz.

2. Speed = Frequency × Wavelength = $20 \times 200 = 4000$ m s^{-1}

3.

(a) Wave B has higher intensity, because its amplitude is larger.

(b) The two waves have the same speed.

4. The beat frequency is $104.3 - 100.1 = 4.2$ Hz

5. The repeat period of the pattern is $2 \times 7 = 14$ s.

6.

(a) 5 m/3 cycles = 1.67 m

(b) 6 cycles in 5 s, so it takes 5/6 = 0.83 s. That is the period of the oscillation.

(c) The inverse of the period, that is 6/5 = 1.2 Hz

(d) The speed: $1.67 \times 1.2 = 2$ m s^{-1}

(e) should be close to 2 m s^{-1} results may vary.

Chapter 2

1.

(a) 3 m

(b) 3 m

(c) 1.5 m

2. The speed of sound is higher in hot air. The round trip (echo time) will be **shorter** in hot air.

3. Sound spreads out of the window (see figure 2.5.) The volume will be higher when facing the window, but the sound will spread just the same.

4.

    (a) time = distance/speed = 3500/343 = 10 s.

    (b) 0.000 01 s. This time is very small compared to 10 s, so we can ignore it, and assume that we see the light instantly.

    (c) distance = time × speed = 5 × 343 = 1700 m (about 1 mile). Hence the rule of thumb, 1 mile for every 5 s!

5. Because longer wavelength will diffract around small objects, so small object will be 'invisible' to long wavelengths. Also, shorter wavelengths spread less than longer wavelengths. That makes it easier to point on the target, and pick up the reflection from the target.

6.

    (a) 69 m s^{-1}

    (b) The speed of sound is about 5 times larger.

    (c) No, because the compressions and rarefactions move much faster. The individual molecules can move back and forth very fast, but on average, their net motion is zero. So, the air molecules transmit the pressure by bumping into each other. Like people making a 'wave' in a stadium

Chapter 3

1. 5000 Hz means that we have 5000 positive and 5000 negative peaks in 1 s. So, we need at least 10 000 points.

2.

    (a) No. It has more than one frequency.

    (b) Yes. It would look similar to figure 1.9, with a beat frequency of 10 Hz.

    (c) Two vertical lines: one at 100 Hz, 5 units high, and one at 110 Hz, 2 units high.

    (d) No. The amplitudes remain the same, 5 and 2 Volts, respectively.

3.

    (a) Neither one is a pure tone, because each has more than one frequency.

    (b) From figure 3.15 we see that the waveform of the quena flute remains steady for a longer time, therefore the amplitude spectrum of the flute (shown in figure 3.16) will be more or less unchanged over that time interval (the 'sustain' part in figure 3.15.) Also, the 'attack' is more gradual for the flute.

4.

    (a) Figures 3.2 and 3.4 represent pure tones.

    (b) figure 3.18 (the square wave) contains the largest number of frequency components.

5.

    (a) The flute, because it does not have a pronounced attack stage, and has a steady sustain stage. The guitar builds up abruptly, and dies out slowly. If played in reverse, it would build up slowly, and die out abruptly.

(b) The sounds 'k' and 't' form by a sudden release of pressure (these sounds are called 'plosives'.) In a way, they have a pronounced attack stage. The sound 'm' is essentially a hum from the nose cavity (it is called a 'nasal') and has no significant attack stage.

(c) Examples: tot, pop, Bob, dad. These words contain plosive consonants.

6. Plucking a guitar string gives a certain amount of vibrational energy to the string, which is slowly converted to sound. The peaks diminish as this energy is being spent. In the case of the flute, we are continuously providing energy, by blowing air into the instrument.

Chapter 4

1.

(a) Power is a characteristic of the source that produces the sound, like a loudspeaker, and is measured in Watts. The intensity relates to how much power crosses a surface of 1 m^2, and it is measured in W m^{-2}. See section 1.5.

(b) The intensity is measured in W m^{-2}. The intensity level compares the intensity to a reference intensity and is measured in dB. See section 4.3.

(c) No, because loudness compares two tones of different frequency. Loudness level compares two tones of the _same_ frequency. See end of section 4.7.

2.

(a) No. The dB is essentially 10 × the logarithm of the intensity ratio (listed in the middle column of table 4.1). So, the intensity of the whisper is 10^3 and 10 × log (10^3) = 10 × 3 = 30 dB. Similarly, for the lawnmower 10 × log (10^9) = 10 × 9 = 90 dB. What we can add are the intensity ratios of the whisper and lawnmower: $10^3 + 10^9$ = 1 000 001 000. The corresponding dB value is 10 × log (1 000 001 000) = 10 × 9.000 0004 = 90.000 04 dB which is essentially equal to the intensity level of the lawnmower in dB, which is 90 dB. In summary: when we have two sources, we add their intensities not their dB. As a general rule, we do not take the sum of dB values. We can take the difference between dB values, as explained in section 4.4.

(b) To get the same intensity level as a jet engine (intensity ratio one trillion) we need 1000 lawnmowers (intensity ratio one billion each).

(c) Each decade of dB means an increase by a factor of 10. There are 4 decades in 40 dB, therefore 10 × 10 × 10 × 10 = 10 000 or 10^4.

(d) The intensity ratio of 10 lawnmowers is 10 × one billion = 10 billion = 10^{10}. To find the intensity level in dB, we multiply the exponent (i.e., 10) times 10. The result is 100 dB.

3.

(a) The two sources will emit the same amount of power.

(b) From table 4.2 we see that at 30 Hz the threshold of hearing is 60 dB, therefore, at this intensity level the 30 Hz tone is not audible at all! The threshold for the 1000 Hz is 0 dB, therefore the tone is audible.

(c) No, they are not the same. The intensity level refers to the rate at which the sound energy is received, whether audible or not, and is a physically measurable quantity. The loudness level refers to our perception of the sound.

(d) The two sources will sound equally loud.

(e) From table 4.4 we see that we need to increase the intensity level by 10 dB to double the loudness. Note that an increase of 10 dB in intensity level means a 10-fold increase in the intensity ratio.

4.

(a) No, because the difference in frequency is less than 3 Hz, which is less than the just noticeable difference.

(b) Yes, because of the beats. The intensity will go through cycles every 0.5 s (because the beat frequency is $1002 - 1000 = 2$ Hz). Remember that the frequency is the inverse of the period of the cycle.

(c) In part (a) the tones are not played simultaneously. In part (b) they are played simultaneously, hence the beats.

5.

(a) The 30 Hz tone is below the threshold of hearing (see table 4.2.)

(b) The 250 Hz tone is audible. The 30 Hz tone is not audible at all.

(c) The two tones have the same sound pressure level, 20 dB.

6.

(a) We can hear the clicking, because the sound transmits through our head bones to the inner ear. Very little sound transmits to the air if the mouth is closed. Some headphones use this so-called bone conduction.

(b) When we record our voice, what we capture is the sound transmitted through the air. When we hear our own voice, we hear the sound that transmits to our inner ear through our bones, in addition to what transmits to the air.

Chapter 5

1.

(a) The heptatonic scale uses seven notes in one octave, the chromatic scale uses all of the 12 notes in the octave.

(b) Using the equal temperament scale, we can transpose to any key, without having to retune.

2.

(a) C-major scale.

(b) For the C-major scale, the third is E, the fourth is F, and fifth is G.

(c) The A-minor scale.

(d) The A-minor third is C, the fourth is D, and fifth is E.

3.

(a) The present standard is sharper (higher frequency) than the baroque standard.
(b) The ratio of the frequencies is 440/415 = 1.060
(c) One semitone interval in the equal temperament scale is 1.059 46 (see table 5.1.)
(d) They are about 1 semitone apart.

4.

(a) The A-major scale is: A, B, C♯, D, E, F♯, and G♯
(b) The natural C-minor scale is: C, D, E♭, F, G, A♭, and B♭

5. We could, but the modes would be redundant. From figure 5.5 we see that the Aeolian mode is the A-minor. So, starting from A and playing the white keys, we would reproduce the same modes as starting from C.

6.

(a) The scale will have will have whole tone steps only. Which means (see figure 5.1) C, D, E, F♯, G♯, and A♯. So, starting, form C4 (the middle C), we play the first three white keys, and then the three black keys. The frequencies corresponding to these notes are listed in appendix C.

7.

(a) The minor third of A4 is C5.
(b) From table 5.3 we see that the frequency ratio of A4 to C4 is 27:16. So the frequency of C4 = 440 × (16/27) = 260.7 Hz. The frequency of C5 is one octave higher than the frequency of C4. So, in the Pythagorean tuning, the frequency of C5 = 2 × 260.7 = 521.5 Hz.
(c) In the same way, from table 5.4 the ratio is 5:3. The frequency of C4 = 440 × (3/5) = 264 Hz. So, in the just scale, the frequency of C5 = 2 × 264 = 528 Hz.
(d) In the equal temperament scale, C5 is 3 semitone intervals above A4. Each semitone interval is 1.059 46. So, in the equal temperament scale, the frequency of C5 = 440 × (1.059 46)^3 = 523.25 Hz, as listed in appendix C.
(e) The fifth is the same for major and minor scale. There is no minor fifth in the heptatonic scale.

Chapter 6

1.

(a) The longest wavelength (the fundamental) in the open pipe is two times the length of the pipe, that is 1.2 m.
(b) The longest wavelength in the clamped string is two times the length of the string, that is, 1.2 m.
(c) No. For a given wavelength, the frequency is determined by the speed of sound in the medium, by the relation frequency = speed/wavelength.

In the pipe, the speed of sound is about 343 m s^{-1}. In the string, the speed of sound depends on the tension of the string and the properties of the string. This is why in a guitar, although the strings have equal open length, they produce different tones.

2.

(a) The frequency of the overtones will be integer multiples of 50 Hz; so, their frequencies are: 100, 150, and 200 Hz.

(b) For the pipe closed at one end the frequencies are odd multiples of the fundamental, so the frequencies are: 150, 250, and 350 Hz.

(c) The open pipe is longer. The length of the open pipe is one-half the wavelength of the fundamental. The length of the closed pipe is one-fourth the wavelength of the fundamental. The wavelength of the fundamental is the same for both pipes in this case, therefore, the open pipe is longer.

3.

(a) The fundamental is 100 Hz

(b) The overtone frequencies are 200, 300, 400 Hz and so on.

(c) The overtones are harmonic because they are integer multiples of the fundamental.

(d) The frequency of the second harmonic is 200 Hz.

(e) The second overtone frequency is 300 Hz.

4.

(a) The overtones are not harmonic because they are not integer multiples of the fundamental.

(b) This is not a pure tone because it contains more than one frequency.

(c) All the frequencies, i.e., 100, 202, 303, 404,...Hz, are partials of this tone.

5.

(a) Using frequency = (n × 343)/(2 × length of pipe) and n = 1, 2, 3, we get 71.5, 143, 214.5 Hz etc.

(b) It should be half as long. Length = 1.2 m.

(c) It should be half as long. Length = 1.2 m.

(d) From part (a) the frequency of the fundamental is 71.5 Hz. A pipe closed at one end produces odd harmonics (3,5,7,...). So, the first overtone is 3 × 71.5 = 214.4 Hz, the second is 357.5 Hz and the third overtone is 500.5 Hz.

6.

(a) The second harmonic has twice the frequency of the fundamental, so the interval of the second harmonic is the octave.

(b) The third harmonic has three times the frequency of the fundamental, so it is a fifth above the octave. For instance, if the fundamental has frequency of 100 Hz, the octave is at 200 Hz, and the third harmonic is 300 Hz. So, the interval between the second and third harmonic is 300/200 = 3:2 which means one fifth above the octave.

(c) Two octaves and one major third. The fourth harmonic is two octaves above the fundamental, the fifth harmonic is five times the fundamental, so it is the third above the fourth harmonic. For instance, if the frequency of the fundamental is 100 Hz, the fourth harmonic is at 400 Hz and the fifth at 500 Hz, and the interval between the fourth and fifth harmonic is 500/400 = 5:4, which is the interval of the major third.

7.

(a) The ear canal behaves like a pipe closed at one end. Fundamental frequency = 343/(4 × 0.03) = 2585 Hz.

(b) Sound frequencies around 2585 Hz resonate in the ear canal, so, they sound louder.

8.

(a) Same as in air. Depends only on length of string.

(b) Same as in air. Depends only on speed of wave *in the string* and length of string.

(c) The two frequencies would be equal.

(d) Twice the length of the pipe, 2 m.

(e) Here we need to use the speed of sound in helium, rather than air. So, from section 6.6, the frequency is 1000/2 = 500 Hz.

(f) The two frequencies would be equal.

Chapter 7

1.

(a) According to Ohm's Law, the lower impedance (5 Ohms) draws more current from the source.

(b) Power is the product of voltage times current. As the voltage is constant, the load that draws more current (the 5 Ohms) draws more power.

(c) A thick wire has very low impedance, and will draw the highest current (and power) the voltage source can afford. In this case we are 'shorting' the source, which usually means that either the wire will burn out or the source will blow a fuse, or be damaged!

2.

(a) Yes, in an AC circuit the direction of the current alternates.

(b) Yes, in an AC circuit the voltage alternates.

(c) No. That would require a way for the electrons to speed-up rather than slowing-down as they travel through the resistor wire. Note however, that according to Ohm's law: voltage = resistance × current. In voltage sources the current and voltage have opposite signs; the current flows *through* the source from the negative to the positive terminal. So, if a circuit element has negative resistance, it is a voltage source.

(d) A thick short wire. It will dissipate a lot of power. See problem 1(c).

(e) No. That would mean that the resistor generates power, and that is what a voltage source does.

3.

(a) A cardioid microphone would be more suitable for a singer in a band, because it can pick up more sound from the singer, less from the other instruments/singers, but also allow the artist some motion.

(b) An omnidirectional microphone would be more suitable for a theater stage because it can pick up sound from all directions.

(c) A cardioid or a shotgun microphone.

4.

(a) There are 2 decades in 20 dB, therefore the output is 10^2 times larger than the input, or 100 Watts.

(b) It would take a total of 200 Watts: 100 W for sound output and 100 W lost as heat.

(c) Of the 100 W fed to the loudspeaker, the sound output will be 10 W and 90 W will be lost as heat. So, of the total 200 Watts supplied to the amplifier, only 10 W will become sound. Not very efficient!

(d) A 4 Ohm load draws more power than an 8 Ohm load.

5. Refer to section 7.6 and figure 7.7.

(a) When the volume setting is low we are still in the linear part of the response curve, therefore the non-linearity is not noticeable.

(b) When the volume setting is high we are beyond the linear part, and the effects of non-linearity are noticeable as harmonic distortion.

(c) When the volume setting is high, the non-linearity will increase the total harmonic distortion.

(d) Harmonic distortion means that multiples of the fundamental frequency will appear. For high frequencies, say 10 000 Hz, the harmonics, are 20 000, 30 000 and so on, which are mostly or totally outside the audible range. For lower frequencies, say 100 Hz, quite a few of the harmonics (200, 300, 400, 500, and so on) are in the audible range. Therefore, the nonlinearity will affect mainly the bass, and midrange.

6.

(a) There are 10 000 cm^2 in 1 m^2.

(b) Intensity is power/area, or 0.002 W/0.0001 m^2 = 20 W m^{-2}.

(c) This intensity is above the threshold of pain. Conclusion: when using earphones, *the volume must be kept low*.

7.

(a) A pure sinusoid is positive for half a cycle, and negative for half a cycle, in a symmetrical way. So, the average is zero.

(b) No, because the average is zero no matter what the amplitude is.

(c) The average of the voltage squared.

(d) The power will increase by a factor of 4.

8.

    (a) Power = Voltage × current.

        For the loudspeaker, the current must be in the order of 1 A.

        For the earphone, the current must be in the order of 10 mA.

    (b) Voltage = Impedance × Current.

        For the loudspeaker, the impedance must be in the order of 10 Ohms.

        For the earphone, the impedance must be in the order of 100 Ohms.

Chapter 8

1.

    (a) The decimal number 6 in binary is 110.

    (b) The binary number 0101 in decimal is 5.

2. The voltage of the wall outlets is analog, because it takes all values between −170 and +170 Volts (see footnote in section 8.2 for voltage values).

3. All of the input/output ports shown are for analog signals. Some sound cards have a joystick port, which could also be used for MIDI. This is a digital port. The USB is a digital port.

4. Digital computers (there are analog computers as well, but not so common) are based on transistors working essentially as switches. As a switch the transistor can have only two states: ON or OFF. This matches the number of digits used in the binary system: 0 or 1.

5.

    (a) It works with 4-bit words, but the binary representation would be: 0000 (for 0 V); 0001 (for 1 V); 0010 (for 2 V) 0011 (for 3 V). The first two digits are always 0, and this is inefficient.

    (b) Yes. The binary representation would be 00 (for 0 V); 01 (for 1 V); 10 (for 2 V) and 11 (for 3 V.) Here we used all the numbers that can be represented by 2 bits, and that is efficient.

    (c) No, because we cannot represent 4 levels with 1-bit, which can only represent 0 or 1.

6. A digital signal is a signal used by a computer to do arithmetic and logical operations. The signal has two states '0' and '1.' A binary signal has two levels. For example, we can use a binary communication code using light signals: Light On = 1 and Light Off = 0. In short, if we use binary for something, like numbering the pages of a book, that does not necessarily imply that we are dealing with a computer signal.

7.

    (a) 4/2 = 2 kHz.

    (b) No, because we would miss all the fundamentals and harmonics over 2 kHz, which means poor quality.

    (c) The music comes through the 'data' or WiFi, which have higher bandwidths.

Chapter 9

1. We use 343 for the speed of sound, and the formula time = distance/speed.
   (a) It will take 5/343 = 0.015 s.
   (b) Round trip is 10 + 10 = 20 m. It takes 20/343 = 0.058 s.
   (c) There are two reflections in each round-trip. At each reflection the sound is reduced by a factor of (½) therefore, in one round-trip the sound is reduced to (½) × (½) = ¼ and ¾ or 75% is lost.
   (d) Here the sound intensity is reduced by a factor of 0.9 at each reflection, therefore 0.9 × 0.9 = 0.81 or the intensity is reduced to 81%, and 19% is lost.
   (e) Recall that the reverberation time is the time it takes for the sound level to drop by 60 dB (a factor of one million). The material that absorbs 10% will give longer reverberation time, because it allows more round-trips before the level is reduced to one millionth.
2. We use 343 for the speed of sound, and the formula time = distance/speed.
   (a) Round-trip distance is 34 + 34 = 68. So, it will take 68/343 = 0.2 s.
   (b) Ignoring absorption by air, after one round-trip the level is reduced to 0.9 × 0.9 = 0.81 or 81% and 19% is lost.
   (c) In the small hall of the previous problem, with 10% absorption, we lose 19% every 0.058 s (the round-trip time), while in the larger hall we lose 19% every 0.2 s (the round-trip time). In other words, we are losing energy at a faster rate in the small hall, which makes the reverberation time shorter. Conclusion: in larger halls, there is longer time between reflections, therefore, the reverberation time is longer.
   (d) A small hall is more likely to have 'intimacy' because all the distances involved are smaller, therefore there is very short time-lag between arrivals of the reflections.
3.
   (a) Using $n = 1$ in the equation for a pipe closed at both ends:

   frequency = (n × 343)/(2 × length of pipe)

   we find:
       28.5 Hz for the 6 m dimension;
       34.3 Hz for the 5 m dimension; and
       57.2 Hz for the 3 m dimension.
   (b) For all three dimensions the frequency of the first mode is 38.1 Hz.
   (c) Room A, because all three dimensions are different. In room B, all three dimensions are the same. It is a cube.
   (d) Room B will resonate very strongly at 38.1 Hz, so room A is better because the resonances are spread out. This tells us that rooms with rectangular floors are better than rooms with square floors.
   (e) Yes, they are the first, second and fourth modes of figure 9.7. The third mode in the figure must come from bouncing between 6 m and 5 m walls. Note that a pipe closed at both ends is not a bad model. We found two of the three lowest modes correctly.

4. Using the model of the pipe closed at both ends, and proceeding as in problem 3, we find that a 1 m × 1 m × 3 m space will have (among others) two overlapping resonances at 171.5 Hz along the 1 m dimension, and 52.7 Hz, along the 3 m dimension. These low frequency resonances add a pleasant 'fullness' to the voice.

Chapter 10

1. Magnetic tapes are re-writable, less susceptible to damage, have in some cases more storage capacity and are more compact.
2. CDs have less noise, they do not deteriorate (no mechanical contact to the surface) and, depending on the format used (for instance, MP3), can have much higher storage capacity.
3. Semiconductor storage devices have no moving parts (less susceptible to failure), are more compact and have much higher storage capacity.
4.

    (a) 10 000 positive peaks in 1 s.
    (b) 10 000 edges are needed to record the waveform for 1 s.
    (c) Two full turns are needed to fit 10 000 edges.
    (d) Two full turns per second. One way to increase the storage capacity would be to have the edges closer by making the size of the pits smaller, but this would require focusing the laser spot more tightly, which has its own challenges.

5.

    (a) For MP3 we have 1 megabyte per minute. So, we need 2 megabytes.
    (b) One byte is 8 bits, so we need $8 \times 2 = 16$ megabits.
    (c) We have 16 000 000 bits in 1 cm^2. This means a matrix of 4000 bits on each side, or 4000 bits per cm. This translates to a spacing of 0.01/4000 = 2.5 millionths of a meter (2.5 microns.)
       This is 20 times smaller than the thickness of a human hair.

Chapter 11

1.

    (a) They all have vibrators.
    (b) Guitar: string; clarinet: air column in the bore; cymbal: the disk itself; drum: the drum head.

2. Guitar: box; clarinet: the open end and holes; cymbal: the disk itself; drum: the drum head.
3. The guitar and the flute produce harmonic overtones, that is, integer multiples of the fundamental. The cymbal, like all two-dimensional surfaces does not produce harmonic overtones. The drum does not produce harmonic overtones.
4.

    (a) The scale length is the distance between the bridge (or saddle in the case of the guitar) and the nut.

(b) The double bass can play the lowest notes (lower frequency and longer wavelength).

(c) Recall that the wavelength of the fundamental in a clamped string is twice the length of the string. In normal tuning, this means that a longer wavelength is generally required for lower notes. So, the double bass has largest scale length.

(d) The violin has the widest tonal range.

5. The explanation is that the flute is an open pipe while the clarinet is a pipe closed at one end. As discussed in section 6.8, pipes closed at one end can be shorter than open instruments, and produce the same frequency.

6. One reason is that if a string is loose, it rattles around. This makes the string difficult to play, especially in bowed instruments.

Another reason is that loose strings do not store much energy, essentially for the same reason that one cannot bounce high from a very soft mattress or a soft trampoline. If the string is loose, the sounds become weaker and last a short time. Also, for instruments like the hammer dulcimer, the mallets have to *bounce* off the string really quickly, and loose strings will not do that. The same applies to the hammers in pianos (section 12.3.3.)

Also, if the strings are loose, the higher frequency modes become weak and the notes sound dull.

You can try tuning-down the first string in a violin or a guitar by one octave, and make your observations.

Chapter 12

1.

(a) They all do.

(b) Pipe organ: air reeds for flue pipes, metal reed for reed pipes;
  piano: strings;
  synthesizer: electronic oscillator;
  human voice: vocal folds.

2. Pipe organ: the pipes; piano: the soundboard; human voice: pharynx, oral and nasal cavities.

3.

(a) No, the vibrational frequency is determined mainly by the mass and tension of the vocal folds.

(b) The resonances in the pharynx, oral and nasal cavities. As discussed in section 6.6, the mode frequencies are determined by the shape of the resonator and on the speed. Since the speed of sound in helium is higher than air, we expect the frequencies to increase. So, we hear the fundamental from the vocal folds unchanged, and all the resonances at higher pitch.

4. We need to consider the end effects, meaning add a correction of $0.6\,r$, where r is the radius of the pipe. Suppose that the pipe radius is 0.12 m. The correction for each end is $0.6 \times 0.12 = 0.072$ m. So, the corrected length is $2.44 + 0.14 = 2.58$ m and the corrected frequency is:

$$\text{frequency} = 343/(2 \times 2.58) = 66.5 \text{ Hz},$$

which is closer to 65.4 Hz.

5.

    (a) Using frequency $= (n \times 343)/(2 \times \text{length of pipe})$ and $n = 1$ we find 42.8 Hz.

    (b) Proceeding as in part (a) we find 64.3 Hz.

    (c) $64.3/42.8 = 1.5$.

    (d) The harmonics of pipe A must be *integer multiples* of 42.8 Hz. 64.3 is not an integer multiple of 42.8.

    (e) $4/2.666 = 1.5$, the same as part (c.) This is so because the frequency is determined by the inverse of the length of the pipe. So, the frequency ratio equals the inverse of the length ratio.

    (f) The ratio $1.5 = 3:2$ is the perfect fifth.

    (g) Pipe Y will produce the fifth of C4, which is G4.

    (h) They should remain the same, because when forming ratios, the units cancel out.

    (i) The note will be G.

    (j) We will be playing in fifths.